'Unlike T. S. Eliot, who wrote that humankind cannot bear too much reality, Dawkins wants us to get real about the universe in which brief lives are set, because reality is liberating ... This is the best book of sermons I have read for years. So please go on preaching to us, Reverend Dawkins, and don't mind the things they throw at you. After all, prophets always get stoned'

Richard Holloway, Bishop of Edinburgh

Richard Dawkins FRS is the Charles Simonyi Professor of the Public Understanding of Science at Oxford, and a Fellow of New College. He was born in Kenya, and read Zoology at Oxford, and did his doctorate under the Nobel Prizewinner Niko Tinbergen. His awards include the International Cosmos Prize for 1997 and the Kistler Prize for 2001. He has written six books since *The Selfish Gene* (1976), has Honorary Doctorates in both science and literature, and is a Fellow of the Royal Society as well as of the Royal Society of Literature. He lives in Oxford with his wife, the actress and artist Lalla Ward.

Latha Menon is an editorial consultant. She was born in India, read physics at Oxford and undertook research in astrophysics before entering publishing 18 years ago. Formerly Executive Editor of Microsoft Encarta Encyclopedia, World English Edition, her interests lie in the conceptual mapping of knowledge and the thoughtful dissemination of learning.

By Richard Dawkins

The Selfish Gene
The Extended Phenotype
The Blind Watchmaker
River Out of Eden
Climbing Mount Improbable
Unweaving the Rainbow
A Devil's Chaplain

A Devil's Chaplain

Selected Essays by
Richard Dawkins

Edited by Latha Menon

PHOENIX

A PHOENIX PAPERBACK

First published in Great Britain in 2003
by Weidenfeld & Nicolson
This paperback edition published in 2004
by Phoenix,
an imprint of Orion Books Ltd,
Orion House, 5 Upper St Martin's Lane,
London WC2H 9EA

A CIP catalogue record for this book
is available from the British Library.

ISBN 0 75381 750 0

Typeset by Selwood Systems, Midsomer Norton

Printed and bound in Great Britain by
Clays Ltd, St Ives plc

CONTENTS

For Juliet on her Eighteenth Birthday

AUTHOR'S NOTE

This book constitutes a personal selection, made by the Editor Latha Menon, from among all the articles and lectures, reflections and polemics, book reviews and forewords, tributes and eulogies that I have published (or in a few cases not previously published) over 25 years. There are many themes here, some arising out of Darwinism or science in general, some concerned with morality, some with religion, education, justice, history of science, some just plain personal.

Though I admit to occasional flames of (entirely justified) irritation in my writing, I like to think that the greater part of it is good-humoured, perhaps even humorous. Where there is passion, well, there is much to be passionate about. Where there is anger, I hope it is a controlled anger. Where there is sadness, I hope it never spills over into despair but still looks to the future. But mostly science is, for me, a source of living joy, and I hope it comes through in these pages.

My contribution to the book itself has been to write the pre-ambles to each of the seven sections, reflecting on the essays Latha has chosen and the connections between them. Hers was the difficult task, and I am filled with admiration for the patience with which she read through vastly more of my writings than are here reproduced, and for the skill with which she achieved a subtler balance of them than I thought they possessed. Her own Introduction gives the reasoning behind her choice, and behind her sorting of the essays into seven sections with a carefully crafted running-order within each section. But as for what she had to choose from, the responsibility is, of course, mine.

It is not possible to list all the people who helped with the individual pieces, spread as they are over 25 years. Help with the book itself came from Yan Wong, Christine DeBlase-Ballstadt, Anthony Cheetham, Michael Dover, Laura van Dam and Catherine Bradley. My gratitude to Charles Simonyi is unabated. And my wife Lalla Ward continues to lend her encouragement, her advice and her fine-tuned ear for the music of language.

Richard Dawkins

EDITOR'S INTRODUCTION

It took quite a while for me to get round to reading *The Selfish Gene*. My love had been for the elegance, the philosophical profundity, the exquisite simplicity of the world as revealed by physics. Chemistry seemed messy, and as for biology – well, my brief acquaintance with it from school had suggested a dry field, full of dull collections of facts, much learning by rote, and little in the way of organizational principles. How wrong I was. Like many, I had thought I understood evolution, but it was through the books of Richard Dawkins in particular that I was introduced to the astonishing depth and grandeur of Darwin's (and Wallace's) idea, to its astounding explanatory power and its profound implications for ourselves and our view of the world. The narrow domestic walls that habit, tradition and prejudice had erected between the fields of science in my mind fell away.

I was delighted, therefore, to be able to repay the debt in some small measure when I was asked by the publishers to put together this collection of Richard's writings. Richard is an academic scientist, but this volume does not include his academic papers. Instead it brings together a number of his shorter articles and columns intended for a wider audience. The task was not an easy one. The composing of this volume has involved some difficult choices and has sadly entailed leaving behind much which must await a future collection. In selecting the pieces included here, I have sought to reflect the range of Richard's interests and concerns, and something of his life too; indeed, almost inevitably, the volume contains an autobiographical element. It is divided into seven sections, moving broadly from science, through memes and religion, to people and memories. The first six sections contain mixtures of pieces of varying lengths and moods, written in different contexts.

There is plenty here, of course, on evolution, and more generally on the nature of science, on its unique ability to seek out truth, contrasted with the muddled thinking of New Age mysticism and spirituality, the superficially more impressive 'metatwaddle' of postmodernism, and the closed, authoritarian, faith-based beliefs of revealed religion. This would not be a representative volume without some of Richard's writings on religion. I have an especially pertinent personal reason for sharing the urgency and passion of his words on the subject: I was born in India – that country which has been so dragged back by its superstitious baggage, where religious labels have been used to such widespread and horrific effect.

So much for the necessary and principled stand. Being a scientist and rationalist does not mean a life of soulless grind, of misery and meaninglessness, but one that is immensely more enriched, more precious. Gathered here, too, then, is a selection of warm memories – of an African childhood, of inspiring mentors, of departed friends, much loved. And books and love of learning weave their way throughout the whole, with forewords, reviews and critical commentaries (including a section on the works of the late Stephen J. Gould).

The final section, 'A Prayer for My Daughter', in many ways sums up the key themes of the volume. It expresses an earnest hope that future generations will continue to strive for an understanding of the natural world through reason and based on evidence. It is a passionate plea against the tyranny of mind-numbing belief systems.

My main task has been the selection and arrangement of Richard's writings. The articles appear much as they did in their original form, with occasional deletions and minor word-changes to fit the context of the collection, and the addition of further explanatory footnotes. Richard himself has been a model of patience and generosity throughout the preparation of the volume, as well as a constant inspiration. My thanks also go to Lalla Ward for her valuable comments and suggestions, Christine DeBlase-Ballstadt for her assistance with the textual material, and Michael Dover and Laura van Dam for their encouragement and support for the project.

A final word. For me as editor, working on this collection has been a particularly special experience, so closely do my own views accord with those of the author on many things. Above all, this volume is about the richness of the world when viewed in the light of scientific understanding. Science reveals a reality wondrous beyond the imaginings of tradition. Look again at that entangled bank.

L.M.

1

SCIENCE AND SENSIBILITY

The first essay in this volume, **A Devil's Chaplain** (1.1), has not previously been published. The title, borrowed by the book, is explained in the essay itself. The second essay, **What is True?** (1.2), was my contribution to a symposium of that name, in *Forbes ASAP* magazine. Scientists tend to take a robust view of truth and are impatient of philosophical equivocation over its reality or importance. It's hard enough coaxing nature to give up her truths, without spectators and hangers-on strewing gratuitous obstacles in our way. My essay argues that we should at least be consistent. Truths about everyday life are just as much – or as little – open to philosophical doubt as scientific truths. Let us shun double standards.

At times I fear turning into a double standards bore. It started in childhood when my first hero, Doctor Dolittle (he returned irresistibly to mind when I read the *Naturalist's Voyage* of my adult hero, Charles Darwin), raised my consciousness, to borrow a useful piece of feminist jargon, about our treatment of animals. Non-human animals I should say, for, of course, we are animals. The moral philosopher most justly credited with raising today's consciousness in this direction is Peter Singer, lately moved from Australia to Princeton. His *The Great Ape Project* aims towards granting the other great apes, as near as is practically possible, civil rights equivalent to those enjoyed by the human great ape. When you stop and ask yourself *why* this seems so immediately ridiculous, the harder you think, the less ridiculous it seems. Cheap cracks like 'I suppose you'll need reinforced ballot-boxes for gorillas, then?' are soon dispatched: we give rights, but not the vote, to children, lunatics and Members of the House of Lords. The biggest objection to the GAP is 'Where will it all end? Rights for oysters?' (Bertrand Russell's quip, in a similar context). Where do you draw the line? **Gaps in the Mind** (1.3), my own contribution to the

GAP book, uses an evolutionary argument to show that we should not be in the business of drawing lines in the first place. There's no law of nature that says boundaries have to be clear-cut.

In December 2000 I was among those invited by David Miliband MP, then Head of the Prime Minister's Policy Unit and now Minister for School Standards, to write a memo on a particular subject for Tony Blair to read over the Christmas holiday. My brief was **Science, Genetics, Risk and Ethics** (1.4) and I reproduce my (previously unpublished) contribution here (eliminating Risk and some other passages to avoid overlap with other essays).

Any proposal to curtail, in the smallest degree, the right of trial by jury is greeted with wails of affront. On the three occasions when I have been called to serve on a jury, the experience proved disagreeable and disillusioning. Much later, two grotesquely over-publicized trials in the United States prompted me to think through a central reason for my distrust of the jury system, and to write it down as **Trial By Jury** (1.5).

Crystals are first out of the box of tricks toted by psychics, mystics, mediums and other charlatans. My purpose in the next article was to explain the real magic of crystals to the readers of a London newspaper, the *Sunday Telegraph*. At one time it was only the low-grade tabloid newspapers that encouraged popular superstitions like crystal-gazing or astrology. Nowadays some up-market newspapers, including the *Telegraph*, have dumbed down to the extent of printing a regular astrology column, which is why I accepted their invitation to write **Crystalline Truth and Crystal Balls** (1.6).

A more intellectual species of charlatan is the target of the next essay, **Postmodernism Disrobed** (1.7). Dawkins' Law of the Conservation of Difficulty states that obscurantism in an academic subject expands to fill the vacuum of its intrinsic simplicity. Physics is a genuinely difficult and profound subject, so physicists need to – and do – work hard to make their language as simple as possible ('but no simpler,' rightly insisted Einstein). Other academics – some would point the finger at continental schools of literary criticism and social science – suffer from what Peter Medawar (I think) called Physics Envy. They want to be thought profound, but their subject is actually rather easy and shallow, so they have to language it up to redress the balance. The physicist Alan Sokal perpetrated a blissfully funny hoax on the Editorial 'Collective' (what else?) of a particularly pretentious journal of social studies. Afterwards, together with his colleague Jean Bricmont, he published a book, *Intellectual Impostures*, ably documenting this epidemic of *Fashionable Nonsense* (as their book was retitled in the United States). 'Postmodernism Disrobed' is my review of this hilarious but disquieting book.

I must add, the fact that the word 'postmodernism' occurs in the title given me by the Editors of *Nature* does not imply that I (or they) know what it means. Indeed, it is my belief that it means nothing at all, except in the restricted context of architecture where it originated. I recommend the following practice, whenever anybody uses the word in some other context. Stop them instantly and ask, in a neutral spirit of friendly curiosity, what it means. Never once have I heard anything that even remotely approaches a usable, or even faintly *coherent*, definition. The best you'll get is a nervous titter and something like, 'Yes I agree, it is a terrible word isn't it, but you know what I mean.' Well no, actually, I don't.

As a lifelong teacher, I fret about where we go wrong in education. I hear horror stories almost daily of ambitious parents or ambitious schools ruining the joy of childhood. And it starts wretchedly early. A six-year-old boy receives 'counselling' because he is 'worried' that his performance in mathematics is falling behind. A headmistress summons the parents of a little girl to suggest that she should be sent for external tuition. The parents expostulate that it is the school's job to teach the child. Why is she falling behind? She is falling behind, explains the headmistress patiently, because the parents of all the other children in the class are paying for them to go to external tutors.

It is not just the joy of childhood that is threatened. It is the joy of true education: of reading for the sake of a wonderful book rather than for an exam; of following up a subject because it is fascinating rather than because it is on a syllabus; of watching a great teacher's eyes light up for sheer love of the subject. **The Joy of Living Dangerously: Sanderson of Oundle** (1.8) is an attempt to bring back from the past the spirit of just such a great teacher.

1.1

A Devil's Chaplain

Darwin was less than half joking when he coined the phrase Devil's Chaplain in a letter to his friend Hooker in 1856.

What a book a Devil's Chaplain might write on the clumsy, wasteful, blundering low and horridly cruel works of nature.

A process of trial and error, completely unplanned and on the massive scale of natural selection, can be expected to be clumsy, wasteful and blundering. Of waste there is no doubt. As I have put it before, the racing elegance of cheetahs and gazelles is bought at huge cost in blood and the suffering of countless antecedents on both sides. Clumsy and blundering though the *process* undoubtedly is, its results are opposite. There is nothing clumsy about a swallow; nothing blundering about a shark. What is clumsy and blundering, by the standards of human drawing boards, is the Darwinian algorithm that led to their evolution. As for cruelty, here is Darwin again, in a letter to Asa Gray of 1860:

I cannot persuade myself that a beneficent and omnipotent God would have designedly created the Ichneumonidae with the express intention of their feeding within the living bodies of Caterpillars.

Darwin's French contemporary Jean Henri Fabre described similar behaviour in a digger wasp, *Ammophila*:

It is the general rule that larvae possess a centre of innervation for each segment. This is so in particular with the Grey Worm, the sacrificial victim of the Hairy Ammophila. The Wasp is acquainted with this anatomical secret: she stabs the caterpillar again and again, from end to end, segment by segment, ganglion by ganglion.[1]

Darwin's Ichneumonidae, like Fabre's digger wasps, sting their prey not to kill but to paralyse, so their larvae can feed on fresh (live) meat. As Darwin clearly understood, blindness to suffering is an inherent consequence of natural selection, although on other occasions he tried to play down the cruelty, suggesting that killing bites are mercifully swift. But the Devil's Chaplain would be equally swift to point out that if there is mercy in nature, it is accidental. Nature is neither kind nor cruel but indifferent. Such kindness as may appear emerges from the same imperative as the cruelty. In the words of one of Darwin's most thoughtful successors, George C. Williams[2],

> With what other than condemnation is a person with any moral sense supposed to respond to a system in which the ultimate purpose in life is to be better than your neighbor at getting genes into future generations, in which those successful genes provide the message that instructs the development of the next generation, in which that message is always 'exploit your environment, including your friends and relatives, so as to maximize our genes' success', in which the closest thing to a golden rule is 'don't cheat, unless it is likely to provide a net benefit'?

Bernard Shaw was driven to embrace a confused idea of Lamarckian evolution purely because of Darwinism's moral implications. He wrote, in the Preface to *Back to Methuselah*:

> When its whole significance dawns on you, your heart sinks into a heap of sand within you. There is a hideous fatalism about it, a ghastly and damnable reduction of beauty and intelligence, of strength and purpose, of honor and aspiration.

His Devil's Disciple was an altogether jollier rogue than Darwin's Chaplain. Shaw didn't think of himself as religious, but he had that childlike inability to distinguish what is true from what we'd like to be true. The same kind of thing drives today's populist opposition to evolution[3]:

> The most evolution could produce would be the idea that 'might makes right'. When Hitler exterminated approximately 10 million innocent men, women, and children, he acted in complete agreement with the theory of evolution and in complete disagreement with everything humans know to be right and wrong ... If you teach children that they evolved from monkeys, then they will act like monkeys.

An opposite response to the callousness of natural selection is to exult in it, along with the Social Darwinists and – astonishingly – H. G. Wells. *The New Republic*, where Wells outlines his Darwinian Utopia, contains some blood-chilling lines:[4]

> And how will the New Republic treat the inferior races? How will it deal with the black? … the yellow man? … the Jew? … those swarms of black, and brown, and dirty-white, and yellow people, who do not come into the new needs of efficiency? Well, the world is a world, and not a charitable institution, and I take it they will have to go … And the ethical system of these men of the New Republic, the ethical system which will dominate the world state, will be shaped primarily to favour the procreation of what is fine and efficient and beautiful in humanity – beautiful and strong bodies, clear and powerful minds … And the method that nature has followed hitherto in the shaping of the world, whereby weakness was prevented from propagating weakness … is death … The men of the New Republic … will have an ideal that will make the killing worth the while.

Wells's colleague Julian Huxley downplayed, in effect, the pessimism of the Devil's Chaplain as he tried to build an ethical system on what he saw as evolution's progressive aspects. His 'Progress, Biological and Other', the first of his *Essays of a Biologist*,[5] contains passages that read almost like a call to arms under evolution's banner:

> [man's] face is set in the same direction as the main tide of evolving life, and his highest destiny, the end towards which he has so long perceived that he must strive, is to extend to new possibilities the process with which, for all these millions of years, nature has already been busy, to introduce less and less wasteful methods, to accelerate by means of his consciousness what in the past has been the work of blind unconscious forces.

I prefer to stand up with Julian's refreshingly belligerent grandfather T. H. Huxley, agree that natural selection is the dominant force in biological evolution unlike Shaw, admit its unpleasantness unlike Julian, and, unlike Wells, fight against it as a human being. Here is T. H., in his Romanes Lecture in Oxford in 1893, on 'Evolution and Ethics':[6]

> Let us understand, once for all, that the ethical progress of society depends, not on imitating the cosmic process, still less in running away from it, but in combating it.

That is G. C. Williams's recommendation today, and it is mine. I hear the bleak sermon of the Devil's Chaplain as a call to arms. As an academic scientist I am a passionate Darwinian, believing that natural selection is, if not the only driving force in evolution, certainly the only known force capable of producing the illusion of purpose which so strikes all who contemplate nature. But at the same time as I support Darwinism as a scientist, I am a passionate anti-Darwinian when it comes to politics and how we should conduct our human affairs. My previous books, such as *The Selfish Gene* and *The Blind Watchmaker*,[7] extol the inescapable factual correctness of the Devil's Chaplain (had Darwin decided to extend the list of melancholy adjectives in the Chaplain's indictment, he would very probably have chosen both 'selfish' and 'blind'). At the same time I have always held true to the closing words of my first book, 'We, alone on earth, can rebel against the tyranny of the selfish replicators.'

If you seem to smell inconsistency or even contradiction, you are mistaken. There is no inconsistency in favouring Darwinism as an academic scientist while opposing it as a human being; any more than there is inconsistency in explaining cancer as an academic doctor while fighting it as a practising one. For good Darwinian reasons, evolution gave us a brain whose size increased to the point where it became capable of understanding its own provenance, of deploring the moral implications and of fighting against them. Every time we use contraception we demonstrate that brains can thwart Darwinian designs. If, as my wife suggests to me, selfish genes are Frankensteins and all life their monster, it is only we that can complete the fable by turning against our creators. We face an almost exact negation of Bishop Heber's lines, 'Though every prospect pleases, And only man is vile.' Yes, man can be vile too, but we are the only potential island of refuge from the implications of the Devil's Chaplain: from the cruelty, and the clumsy, blundering waste.

For our species, with its unique gift of foresight – product of the simulated virtual-reality we call the human imagination – can plan the very opposite of waste with, if we get it right, a minimum of clumsy blunders. And there is true solace in the blessed gift of understanding, even if *what* we understand is the unwelcome message of the Devil's Chaplain. It is as though the Chaplain matured and offered a second half to the sermon. Yes, says the matured Chaplain, the historic process that caused you to exist is wasteful, cruel and low. But exult in your existence, because that very process has blundered unwittingly on its own negation. Only a small, local negation, to be sure: only one species, and only a minority of the members of that species; but there lies hope.

Exult even more that the clumsy and cruel algorithm of natural selection has generated a machine capable of internalizing the algorithm, setting up a model of itself – and much more – in microcosm inside the human skull. I may have disparaged Julian Huxley in these pages, but he published a poem in 1926 which says something of what I want to say[8] (and a few things that I don't want to say):

> The world of things entered your infant mind
> To populate that crystal cabinet.
> Within its walls the strangest partners met,
> And things turned thoughts did propagate their kind.
> For, once within, corporeal fact could find
> A spirit. Fact and you in mutual debt
> Built there your little microcosm – which yet
> Had hugest tasks to its small self assigned.
>
> Dead men can live there, and converse with stars:
> Equator speaks with pole, and night with day:
> Spirit dissolves the world's material bars –
> A million isolations burn away.
> The Universe can live and work and plan,
> At last made God within the mind of man.

Julian Huxley later wrote, in his *Essays of a Humanist*:[9]

> This earth is one of the rare spots in the cosmos where mind has flowered. Man is a product of nearly three billion years of evolution, in whose person the evolutionary process has at last become conscious of itself and its possibilities. Whether he likes it or not, he is responsible for the whole further evolution of our planet.

Huxley's fellow luminary of the neo-Darwinian synthesis, the great Russian-American geneticist Theodosius Dobzhansky said something similar:[10]

> In giving rise to man, the evolutionary process has, apparently for the first and only time in the history of the Cosmos, become conscious of itself.

So, the Devil's Chaplain might conclude, Stand tall, Bipedal Ape. The shark may outswim you, the cheetah outrun you, the swift outfly you, the capuchin outclimb you, the elephant outpower you, the redwood outlast you. But you have the biggest gifts of all: the gift of understanding the ruthlessly cruel process that gave us all existence; the gift of revulsion against its implications; the gift of foresight – something utterly foreign to the blundering short-term ways of natural selection – and the gift of internalizing the very cosmos.

We are blessed with brains which, if educated and allowed free rein, are capable of modelling the universe, with its physical laws in which the Darwinian algorithm is embedded. As Darwin himself put it, in the famous closing lines of the *Origin of Species*:

> Thus, from the war of nature, from famine and death, the most exalted object which we are capable of conceiving, namely, the production of the higher animals, directly follows. There is grandeur in this view of life, with its several powers, having been originally breathed* into a few forms or into one; and that, whilst this planet has gone cycling on according to the fixed law of gravity, from so simple a beginning endless forms most beautiful and most wonderful have been, and are being, evolved.

There is more than just grandeur in this view of life, bleak and cold though it can seem from under the security blanket of ignorance. There is deep refreshment to be had from standing up full-face into the keen wind of understanding: Yeats's 'Winds that blow through the starry ways'. In another essay, I quote the words of an inspiring teacher, F. W. Sanderson, who urged his pupils to 'live dangerously ...'

---

*In the Second Edition, and all subsequent editions of the *Origin*, the three words 'by the Creator' were interpolated at this point, presumably as a sop to religious sensibilities.

… full of the burning fire of enthusiasm, anarchic, revolutionary, energetic, daemonic, Dionysian, filled to overflowing with the terrific urge to create – such is the life of the man who risks safety and happiness for the sake of growth and happiness.

Safety and happiness would mean being satisfied with easy answers and cheap comforts, living a warm comfortable lie. The daemonic alternative urged by my matured Devil's Chaplain is risky. You stand to lose comforting delusions: you can no longer suck at the pacifier of faith in immortality. To set against that risk, you stand to gain 'growth and happiness'; the joy of knowing that you have grown up, faced up to what existence means; to the fact that it is temporary and all the more precious for it.†

†Note added in proof: I was unaware, when I chose this title, that the BBC had used Darwin's phrase, 'Devil's Chaplain', for an excellent documentary based on Adrian Desmond and James Moore's biography.

What is True?[11]

A little learning is a dangerous thing. This has never struck me as a particularly profound or wise remark,* but it comes into its own in the special case where the little learning is in philosophy (as it often is). A scientist who has the temerity to utter the t-word ('true') is likely to encounter a form of philosophical heckling which goes something like this:

> There is no absolute truth. You are committing an act of personal faith when you claim that the scientific method, including mathematics and logic, is the privileged road to truth. Other cultures might believe that truth is to be found in a rabbit's entrails, or the ravings of a prophet up a pole. It is only your personal faith in science that leads you to favour your brand of truth.

That strand of half-baked philosophy goes by the name of cultural relativism. It is one aspect of the *Fashionable Nonsense* detected by Alan Sokal and Jean Bricmont,[12] or the *Higher Superstition* of Paul Gross and Norman Levitt.[13] The feminist version is ably exposed by Daphne Patai and Noretta Koertge, authors of *Professing Feminism: Cautionary Tales from the Strange World of Women's Studies*:[14]

> Women's Studies students are now being taught that logic is a tool of domination ... the standard norms and methods of scientific inquiry are sexist because they are incompatible with 'women's ways of knowing' ... These 'subjectivist' women see the methods of logic, analysis and abstraction as 'alien territory belonging to men' and 'value intuition as a safer and more fruitful approach to truth'.

How should scientists respond to the allegation that our 'faith' in

*Pope's original is wonderful, but the aphorism doesn't survive isolation from its context.

17

logic and scientific truth is just that – faith – not 'privileged' (favourite in-word) over alternative truths? A minimal response is that science gets results. As I put it in *River Out of Eden*,[15]

> Show me a cultural relativist at 30,000 feet and I'll show you a hypocrite
> … If you are flying to an international congress of anthropologists or
> literary critics, the reason you will probably get there – the reason you
> don't plummet into a ploughed field – is that a lot of Western scientifically
> trained engineers have got their sums right.

Science boosts its claim to truth by its spectacular ability to make matter and energy jump through hoops on command, and to predict what will happen and when.

But is it still just our Western scientific bias to be impressed by accurate prediction; impressed by the power to slingshot rockets around Jupiter to reach Saturn, or intercept and repair the Hubble telescope; impressed by logic itself? Well, let's concede the point and think sociologically, even democratically. Suppose we agree, temporarily, to treat scientific truth as just one truth among many, and lay it alongside all the rival contenders: Trobriand truth, Kikuyu truth, Maori truth, Inuit truth, Navajo truth, Yanomamo truth, !Kung San truth, feminist truth, Islamic truth, Hindu truth. The list is endless – and thereby hangs a revealing observation.

In theory, people could switch allegiance from any one 'truth' to any other if they decide it has greater merit. On what basis might they do so? Why would one change from, say, Kikuyu truth to Navajo truth? Such merit-driven switches are rare. With one crucially important exception. Scientific truth is the only member of the list which regularly persuades converts of its superiority. People are loyal to other belief systems for one reason only: they were brought up that way, and they have never known anything better. When people are lucky enough to be offered the opportunity to vote with their feet, doctors and their kind prosper while witch doctors decline. Even those who do not, or cannot, avail themselves of a scientific education, choose to benefit from the technology that is made possible by the scientific education of others. Admittedly, religious missionaries have successfully claimed converts in great numbers all over the underdeveloped world. But they succeed not because of the merits of their religion but because of the science-based technology for which it is pardonably, but wrongly, given credit.

> Surely the Christian God must be superior to our Juju, because Christ's representatives come bearing rifles, telescopes, chainsaws, radios, almanacs that predict eclipses to the minute, and medicines that work.

So much for cultural relativism. A different type of truth-heckler prefers to drop the name of Karl Popper or (more fashionably) Thomas Kuhn:

> There is no absolute truth. Your scientific truths are merely hypotheses that have so far failed to be falsified, destined to be superseded. At worst, after the next scientific revolution, today's 'truths' will seem quaint and absurd, if not actually false. The best you scientists can hope for is a series of approximations which progressively reduce errors but never eliminate them.

The Popperian heckle partly stems from the accidental fact that philosophers of science are traditionally obsessed with one piece of scientific history: the comparison between Newton's and Einstein's theories of gravitation. It is true that Newton's inverse square law has turned out to be an approximation, a special case of Einstein's more general formula. If this is the only piece of scientific history you know, you might indeed conclude that all apparent truths are mere approximations, fated to be superseded. There is even a quite interesting sense in which all our sensory perceptions – the 'real' things that we 'see with our own eyes' – may be regarded as unfalsified 'hypotheses' about the world, vulnerable to change. This provides a good way to think about illusions such as the Necker Cube.

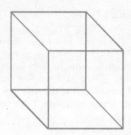

The flat pattern of ink on paper is compatible with two alternative 'hypotheses' of solidity. So we see a solid cube which, after a few seconds, 'flips' to a different cube, then flips back to the first cube, and so on. Perhaps sense data only ever confirm or reject mental 'hypotheses' about what is out there.[16]

Well, that is an interesting theory; so is the philosopher's notion that science proceeds by conjecture and refutation; and so is the analogy between the two. This line of thought – all our percepts are hypothetical models in the brain – might lead us to fear some future blurring of the distinction between reality and illusion in our descendants, whose lives will be even more dominated by computers capable of generating vivid models of their own. Without venturing into the high-tech worlds of virtual reality, we already know that our senses are easily deceived. Conjurors – professional illusionists – can persuade us, if we lack a sceptical foothold in reality, that something supernatural is going on. Indeed, some notorious erstwhile conjurors make a fat living doing exactly that: a living much fatter than they ever enjoyed when they frankly admitted that they were conjurors.* Scientists, alas, are not best equipped to unmask telepathists, mediums and spoon-bending charlatans. This is a job which is best handed over to the professionals, and that means other conjurors. The lesson that conjurors, the honest variety and the impostors, teach us is that an uncritical faith in our own senses is not an infallible guide to truth.

But none of this seems to undermine our ordinary concept of what it means for something to be true. If I am in the witness box, and prosecuting counsel wags his stern finger and demands, 'Is it or is it not true that you were in Chicago on the night of the murder?', I should get pretty short shrift if I said,

> What do you mean by true? The hypothesis that I was in Chicago has not so far been falsified, but it is only a matter of time before we see that it is a mere approximation.

*Performing psychics and mystics, who happily perform in front of scientists, will conveniently plead a headache and refuse to go on if informed that a contingent of professional conjurors is in the front row of the stalls. It is for the same reason that the then Editor of *Nature*, John Maddox, took James 'The Amazing' Randi with him when investigating a suspected case of homeopathic fraud. This caused some resentment at the time, but it was an entirely reasonable decision. Any genuine scientist has nothing to fear from a sceptical conjuror looking over his shoulder.

Or, reverting to the first heckle, I would not expect a jury, even a Bongolese jury, to give a sympathetic hearing to my plea that,

> It is only in your western scientific sense of the word 'in' that I was in Chicago. The Bongolese have a completely different concept of 'in', according to which you are only truly 'in' a place if you are an anointed elder entitled to take snuff from the dried scrotum of a goat.

It is simply true that the Sun is hotter than the Earth, true that the desk on which I am writing is made of wood. These are not hypotheses awaiting falsification; not temporary approximations to an ever-elusive truth; not local truths that might be denied in another culture. And the same can safely be said of many scientific truths, even where we can't see them 'with our own eyes'. It is forever true that DNA is a double helix, true that if you and a chimpanzee (or an octopus or a kangaroo) trace your ancestors back far enough you will eventually hit a shared ancestor. To a pedant, these are still hypotheses which might be falsified tomorrow. But they never will be. Strictly, the truth that there were no human beings in the Jurassic Period is still a conjecture, which could be refuted at any time by the discovery of a single fossil, authentically dated by a battery of radiometric methods. It could happen. Want a bet? Even if they are nominally hypotheses on probation, these statements are true in exactly the same sense as the ordinary truths of everyday life; true in the same sense as it is true that you have a head, and that my desk is wooden. If scientific truth is open to philosophic doubt, it is no more so than common sense truth. Let's at least be even-handed in our philosophical heckling.

A more profound difficulty now arises for our scientific concept of truth. Science is very much not synonymous with common sense. Admittedly, that doughty scientific hero T. H. Huxley said:

> Science is nothing but trained and organized common sense, differing from the latter only as a veteran may differ from a raw recruit: and its methods differ from those of common sense only as far as the guardsman's cut and thrust differ from the manner in which a savage wields his club.

But Huxley was talking about the methods of science, not its conclusions. As Lewis Wolpert emphasized in *The Unnatural Nature of Science*,[17] the conclusions can be disturbingly counter-intuitive.

Quantum theory is counter-intuitive to the point where the physicist sometimes seems to be battling insanity. We are asked to believe that a single quantum behaves like a particle in going through one hole instead of another, but simultaneously behaves like a wave in interfering with a non-existent copy of itself, if another hole is opened through which that non-existent copy *could* have travelled (if it had existed). It gets worse, to the point where some physicists resort to a vast number of parallel but mutually unreachable worlds, which proliferate to accommodate every alternative quantum event; while other physicists, equally desperate, suggest that quantum events are determined retrospectively by our decision to examine their consequences. Quantum theory strikes us as so weird, so defiant of common sense, that even the great Richard Feynman was moved to remark, 'I think I can safely say that nobody understands quantum mechanics.' Yet the many predictions by which quantum theory has been tested stand up, with an accuracy so stupendous that Feynman compared it to measuring the distance between New York and Los Angeles accurately to the width of one human hair. On the basis of these stunningly successful predictions, quantum theory, or some version of it, seems to be as true as anything we know.

Modern physics teaches us that there is more to truth than meets the eye; or than meets the all too limited human mind, evolved as it was to cope with medium-sized objects moving at medium speeds through medium distances in Africa. In the face of these profound and sublime mysteries, the low-grade intellectual poodling of pseudo-philosophical poseurs seems unworthy of adult attention.

Gaps in the Mind[18]

Sir,
You appeal for money to save the gorillas. Very laudable, no doubt. But it doesn't seem to have occurred to you that there are thousands of *human* children suffering on the very same continent of Africa. There'll be time enough to worry about gorillas when we've taken care of every last one of the kiddies. Let's get our priorities right, *please*!

This hypothetical letter could have been written by almost any well-meaning person today. In lampooning it, I don't mean to imply that a good case could not be made for giving human children priority. I expect it could, and also that a good case could be made the other way. I'm only trying to point the finger at the *automatic*, unthinking nature of the speciesist double standard. To many people it is simply self-evident, *without any discussion*, that humans are entitled to special treatment. To see this, consider the following variant on the same letter:

Sir,
You appeal for money to save the gorillas. Very laudable, no doubt. But it doesn't seem to have occurred to you that there are thousands of *aardvarks* suffering on the very same continent of Africa. There'll be time enough to worry about gorillas when we've saved every last one of the aardvarks. Let's get our priorities right, *please*!

This second letter could not fail to provoke the question: What's so special about aardvarks? A good question, and one to which we should require a satisfactory answer before we took the letter seriously. Yet the first letter, I suggest, would not for most people provoke the equivalent question, 'What's so special about humans?' As I said, I don't deny that this question, unlike the aardvark question, very probably has a powerful answer. All that I am criticizing is an unthinking failure to realize in the case of humans that the question even arises.

The speciesist* assumption that lurks here is very simple. Humans are humans and gorillas are animals. There is an unquestioned yawning gulf between them such that the life of a single human child is worth more than the lives of all the gorillas in the world. The 'worth' of an animal's life is just its replacement cost to its owner – or, in the case of a rare species, to humanity. But tie the label *Homo sapiens* even to a tiny piece of insensible, embryonic tissue, and its life suddenly leaps to infinite, uncomputable value.

This way of thinking characterizes what I want to call the discontinuous mind. We'd all agree that a six-foot woman is tall, and a five-foot woman is not. Words like 'tall' and 'short' tempt us to force the world into qualitative classes, but this doesn't mean that the world really is discontinuously distributed. Were you to tell me that a woman is five feet nine inches tall, and ask me to decide whether she should therefore be called tall or not, I'd shrug and say, 'She's five foot nine, doesn't that tell you what you need to know?' But the discontinuous mind, to caricature it a little, would go to court to decide (probably at great expense) whether the woman was tall or short. Indeed, I hardly need to say caricature. For years, South African courts have done a brisk trade adjudicating whether particular individuals of mixed parentage count as white, black or 'coloured'.†

The discontinuous mind is ubiquitous. It is especially influential when it afflicts lawyers and the religious (not only are all judges lawyers; a high proportion of politicians are too, and all politicians have to woo the religious vote). Recently, after giving a public lecture, I was cross-examined by a lawyer in the audience. He brought the full weight of his legal acumen to bear on a nice point of evolution. If species A evolves into a later species B, he reasoned closely, there must come a point when a mother belongs to the old species A and her child belongs to the new species B. Members of

---

*Coined by Richard Ryder and given currency by Peter Singer, the analogy is to racism.
†Thankfully no longer. The apartheid regime is one of history's monuments to the tyranny of the discontinuous mind.

different species cannot interbreed with one another. I put it to you, he went on, that a child could hardly be so different from its parents that it could not interbreed with their kind. So, he wound up triumphantly, isn't this a fatal flaw in the theory of evolution?

But it is we that choose to divide animals up into discontinuous species. On the evolutionary view of life there must have been intermediates, even though, conveniently for our naming rituals, they are today usually extinct. They are not always extinct. The lawyer would be surprised and, I hope, intrigued by so-called 'ring species'. The best-known case is the Herring Gull/Lesser Black-backed Gull ring. In Britain these are clearly distinct species, quite different in colour. Anybody can tell them apart. But if you follow the population of Herring Gulls westward round the North Pole to North America, then via Alaska across Siberia and back to Europe again, you notice a curious fact. The 'Herring Gulls' gradually become less and less like Herring Gulls and more and more like Lesser Black-backed Gulls until it turns out that our European Lesser Black-backed Gulls actually are the other end of a ring that started out as Herring Gulls. At every stage around the ring, the birds are sufficiently similar to their neighbours to interbreed with them. Until, that is, the ends of the continuum are reached, in Europe. At this point the Herring Gull and the Lesser Black-backed Gull never interbreed, although they are linked by a continuous series of interbreeding colleagues all the way round the world. The only thing that is special about ring species like these gulls is that the intermediates are still alive. All pairs of related species are potentially ring species. The intermediates must have lived once. It is just that in most cases they are now dead.

The lawyer, with his trained discontinuous mind, insists on placing individuals firmly in this species or that. He does not allow for the possibility that an individual might lie halfway between two species, or a tenth of the way from species A to species B. Self-styled 'pro-lifers', and others that indulge in footling debates about exactly when in its development a foetus 'becomes human', exhibit the same discontinuous mentality. It is no use telling these people that, depending upon the human charac-teristics that interest you, a foetus can be 'half human' or 'a hundredth human'. 'Human', to the discontinuous mind, is an absolutist concept. There can be no half measures. And from this flows much evil.

The word 'apes' usually means chimpanzees, gorillas, orang utans, gibbons and siamangs. We admit that we are like apes, but we seldom realize that we *are* apes. Our common ancestor with the chimpanzees and gorillas is much more recent than their common ancestor with the Asian apes – the gibbons and orang utans. There is no natural category that includes chimpanzees, gorillas and orangs but excludes humans. The artificiality of the category 'apes', as conventionally taken to exclude humans, is demonstrated by the following diagram. The family tree shows humans to be in the thick of the ape cluster; the artificiality of the conventional category 'ape' is shown by the stippling.

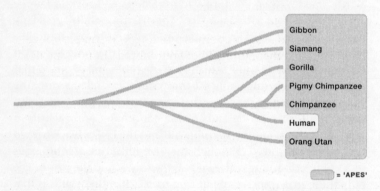

In truth, not only are we apes, we are African apes. The category 'African apes', if you don't arbitrarily exclude humans, is a natural category. The stippled area doesn't have any artificial 'bites' taken out of it:

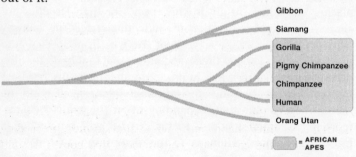

All the African apes that have ever lived, including ourselves, are linked to one another by an unbroken chain of parent-child bonds. The same is true of all animals and plants that have ever lived, but there the distances involved are much greater. Molecular evidence suggests that our common ancestor with chimpanzees lived, in Africa, between 5 and 7 million years ago, say half a million generations ago. This is not long by evolutionary standards.

Happenings are sometimes organized at which thousands of people hold hands and form a human chain, say from coast to coast of the United States, in aid of some cause or charity. Let us imagine setting one up along the equator, across the width of our home continent of Africa. It is a special kind of chain, involving parents and children, and we'll have to play tricks with time in order to imagine it. You stand on the shore of the Indian Ocean in southern Somalia, facing north, and in your left hand you hold the right hand of your mother. In turn she holds the hand of her mother, your grandmother. Your grandmother holds her mother's hand, and so on. The chain wends its way up the beach, into the arid scrubland and westwards on towards the Kenya border.

How far do we have to go until we reach our common ancestor with the chimpanzees? It's a surprisingly short way. Allowing one yard per person, we arrive at the ancestor we share with chimpanzees in under 300 miles. We've hardly started to cross the continent; we're still not halfway to the great Rift Valley. The ancestor is standing well to the east of Mount Kenya, and holding in her hand an entire chain of her lineal descendants, culminating in you standing on the Somali beach.

The daughter that she is holding in her right hand is the one from whom we are descended. Now the arch-ancestress turns eastward to face the coast, and with her left hand grasps her other daughter, the one from whom the chimpanzees are descended (or son, of course, but let's stick to females for convenience). The two sisters are facing one another, and each holding their mother by the hand. Now the second daughter, the chimpanzee ancestress, holds her daughter's hand, and a new chain is formed, proceeding back towards the coast. First cousin faces first cousin, second

27

cousin faces second cousin, and so on. By the time the folded-back chain has reached the coast again, it consists of modern chimpanzees. You are face to face with your chimpanzee cousin, and you are joined to her by an unbroken chain of mothers holding hands with daughters. If you walked up the line like an inspecting general – past *Homo erectus*, *Homo habilis*, perhaps *Australopithecus afarensis* – and down again the other side (the intermediates on the chimpanzee side are unnamed because, as it happens, no fossils have been found), you would nowhere find any sharp discontinuity. Daughters would resemble mothers just as much (or as little) as they always do. Mothers would love daughters, and feel affinity with them, just as they always do. And this hand-in-hand continuum, joining us seamlessly to chimpanzees, is so short that it barely makes it past the hinterland of Africa, the mother continent.

Our chain of African apes in time, doubling back on itself, is in miniature like the ring of gulls in space, except that the intermediates happen to be dead. The point I want to make is that, as far as morality is concerned, it should be incidental that the intermediates are dead. What if they were not? What if a clutch of intermediate types had survived, enough to link us to modern chimpanzees by a chain, not just of hand-holders, but of inter-breeders? Remember the song, 'I've danced with a man, who's danced with a girl, who's danced with the Prince of Wales'? We can't (quite) interbreed with modern chimpanzees, but we'd need only a handful of intermediate types to be able to sing: 'I've bred with a man, who's bred with a woman, who's bred with a chimpanzee.'

It is sheer luck that this handful of intermediates no longer exists. (Good luck from some points of view: for myself, I should love to meet them.) But for this chance, our laws and our morals would be very different. We need only discover a single survivor, say a relict *Australopithecus* in the Budongo Forest, and our precious system of norms and ethics would come crashing about our ears. The boundaries with which we segregate our world would be all shot to pieces. Racism would blur with speciesism in obdurate and vicious confusion. Apartheid, for those that believe in it, would assume a new and perhaps a more urgent import.

But why, a moral philosopher might ask, should this matter to us? Isn't it only the discontinuous mind that wants to erect barriers anyway? So what if, in the continuum of all apes that have lived in Africa, the survivors happen to leave a convenient gap between *Homo* and *Pan*? Surely we should, in any case, not base our treatment of animals on whether or not we can interbreed with them. If we want to justify double standards – if society agrees that people should be treated better than, say, cows (cows may be cooked and eaten, people may not) – there must be better reasons than cousinship. Humans may be taxonomically distant from cows, but isn't it more important that we are brainier? Or [better], following Jeremy Bentham, that humans can suffer more. Or that cows, even if they hate pain as much as humans do (and why on earth should we suppose otherwise?), do not know what is coming to them? Suppose that the octopus lineage had happened to evolve brains and feelings to rival ours. They easily might have done. The mere possibility shows the incidental nature of cousinship. So, the moral philosopher asks, why emphasize the human/chimp continuity?

Yes, in an ideal world we probably should come up with a better reason than cousinship for, say, preferring carnivory to cannibalism. But the melancholy fact is that, at present, society's moral attitudes rest almost entirely on the discontinuous, speciesist imperative.

If somebody succeeded in breeding a chimpanzee/human hybrid, the news would be earth-shattering. Bishops would bleat, lawyers would gloat in anticipation, conservative politicians would thunder, socialists wouldn't know where to put their barricades. The scientist that achieved the feat would be drummed out of common-rooms; denounced in pulpit and gutter press; condemned, perhaps, by an Ayatollah's fatwah. Politics would never be the same again, nor would theology, sociology, psychology or most branches of philosophy. The world that would be so shaken, by such an incidental event as a hybridization, is a speciesist world indeed, dominated by the discontinuous mind.

I have argued that the discontinuous gap between humans and 'apes' that we erect in our minds is regrettable. I have also argued that, in any case, the present position of the hallowed gap is arbitrary, the result of evolutionary accident. If the contingencies of survival and extinction had been different, the gap would be in a different place. Ethical principles that are based upon accidental caprice should not be respected as if cast in stone.

Science, Genetics and Ethics:
Memo for Tony Blair

Senior Ministers (and their Sir Humphreys) could be forgiven for seeing scientists as little more than alternate igniters and quenchers of public panic. If a scientist appears in a newspaper today, it will usually be to pronounce on the dangers of food additives, mobile phones, sunbathing or electricity pylons. I suppose this is inevitable, given the equally forgivable preoccupation of citizens with their own personal safety, and their tendency to hold governments responsible for it. But it casts scientists in a sadly negative role. And it fosters the unfortunate impression that their credentials flow from factual knowledge. What really makes scientists special is less their knowledge than their method of acquiring it – a method that anybody could adopt with advantage.

Even more important, it leaves out the cultural and aesthetic value of science. It is as though one met Picasso and devoted the whole conversation to the dangers of licking one's brush. Or met Bradman* and talked only of the best box protector to put down one's trousers. Science, like painting (and some would say like cricket), has a higher aesthetic. Science can be poetry. Science can be spiritual, even religious in a non-supernatural sense of the word.

In a short memo it is obviously unrealistic to attempt comprehensive coverage of the kind that you will get anyway from civil service briefings. Instead, I thought I would pick out a few isolated topics, vignettes almost, that I find interesting and I hope that you might too. Given more space, I would have mentioned other vignettes (such as nanotechnology, which I suspect we shall be hearing a lot about in the twenty-first century).

*Note to American readers: Sir Donald Bradman (1908–2001) was a cricketer widely regarded, even outside Australia, as the best batsman ever.

Genetics

It is hard to exaggerate the sheer intellectual excitement of post-Watson/Crick genetics. What has happened is that genetics has become a branch of Information Technology. The genetic code is truly digital, in exactly the same sense as computer codes. This is not some vague analogy, it is the literal truth. Moreover, unlike computer codes, the genetic code is universal. Modern computers are built around a number of mutually incompatible machine languages, determined by their processor chips. The genetic code, on the other hand, with a few very minor exceptions, is *identical* in every living creature on this planet, from sulphur bacteria to giant redwood trees, from mushrooms to men. All living creatures, on this planet at least, are the same 'make'.

The consequences are amazing. It means that a software sub-routine (that's exactly what a gene is) can be Copied from one species and Pasted into another species, where it will work exactly as it did in the original species. This is why the famous 'antifreeze' gene, originally evolved by Arctic fish, can save a tomato from frost damage. In the same way, a NASA programmer who wants a neat square root routine for his rocket guidance system might import one from a financial spreadsheet. A square root is a square root is a square root. A program to compute it will serve as well in a space rocket as in a financial projection.

What, then, of the widespread gut hostility, amounting to revulsion, against all such 'transgenic' imports? I suspect that it comes from a pre-Watson/Crick misconception. Surely, the appealing but erroneous reasoning goes, an antifreeze gene from a fish must come with a fishy 'flavour'. Surely some of its fishiness must rub off? Surely it is 'unnatural' to splice a fish gene, which was only ever 'meant' to work in a fish, into the alien environment of a tomato cell? Yet nobody thinks that a square root subroutine carries a 'financial flavour' with it when you paste it into a rocket guidance system. The very idea of 'flavour' in this sense is not just wrong but profoundly and interestingly wrong. It is a cheerful thought, by the way, that most young people today understand computer software far better than their elders, and they should grasp the point instantly. The present Luddism over

SCIENCE, GENETICS AND ETHICS

genetic engineering may die a natural death as the computer-illiterate generation is superseded.

Is there nothing, then, absolutely nothing, in the misgivings of Prince Charles, Lord Melchett and their friends? I wouldn't go that far, although they are certainly muddleheaded.* The square root analogy might be unfair in the following respect. What if it isn't a square root that the rocket guidance program needs, but another function which is not literally *identical* to the financial equivalent? Suppose it is sufficiently similar that the main routine can indeed be borrowed, but it still needs tweaking in detail. In that case, it is possible that the rocket could misfire if we naively import the subroutine raw. Switching back to biology, although genes really are watertight subroutines of digital software, they are *not* watertight in their effects on the development of the organism, for here they interact with their environment, including importantly the environment furnished by other genes. The antifreeze gene might depend, for optimal effect, on an interaction with other genes in the fish. Plonk it down in the foreign genetic climate of a tomato, and it might not work properly unless tweaked (which can be done) to mesh with the existing tomato genes.

What this means is that there is a case to be made on both sides of the argument, and we need to exercise subtle judgement. The genetic engineers are right that we can save time and trouble by climbing on the back of the millions of years of R & D that Darwinian natural selection has put into developing biological antifreeze (or whatever we are seeking). But the doomsayers would also have a point if they softened their stance from emotional gut rejection to a rational plea for rigorous safety testing. No reputable scientist would oppose such a plea. It is rightly routine for all new products, not just genetically engineered ones.

A largely unrecognized danger of the obsessive hysteria surrounding genetically modified foods is crying wolf. I fear that, if the green movement's high-amplitude warnings over GMOs turn

*I explained why in an Open Letter to Prince Charles, *The Observer*, 21 May 2000, http://www.guardian.co.uk/Archive/Article/0,4273,4020558,00.html. See also my article on Lord Melchett's vandalizing of scientific trials of GM crops, *The Observer*, 24 September 2000, http://www.guardian.co.uk/gmdebate/Story/0,2763,372528,00.html.

out to be empty, people will be dangerously disinclined to listen to other and more serious warnings. The evolution of antibiotic resistance among bacteria is a vicious wolf of proven danger. Yet the menacing footfalls of this certain peril are all but drowned out in the caterwauling shrieks over genetically modified foods, whose dangers are speculative at most. To be more precise, genetic modification, like any other kind of modification, is good if you modify in a good direction, bad if you modify in a bad direction. Like domestic breeding, and like natural selection itself, the trick is to introduce the right new DNA software. The realization that software is all it is, written in exactly the same language as the organism's 'own' DNA, should go a long way towards dispelling the gut fears that rule most discussions of GMOs.

I can't leave the subject of gut feelings without a favourite quote from the lamented Carl Sagan. When asked a futurological question, he said that not enough was known to answer it. The questioner pressed him on what he really thought: 'What is your gut feeling?' Sagan's reply is immortal: 'But I try not to think with my gut.' Gut thinking is one of the main problems we have to contend with in public attitudes to science. I shall return to the point under Ethics. Meanwhile, some more remarks on the future of genetics in the twenty-first century, especially in the wake of the Human Genome Project (HGP).

The HGP, which will be completed any time now, is really a twentieth-century accomplishment. It is an outstanding success story, but it has limited scope. We have taken the human hard disk and transcribed every jot and tittle of the 11000101000010000111-style bits of information on it, regardless of what they mean in the software as a whole. The HGP needs to be followed up by a twenty-first-century Human Embryology Project (HEP) which, in effect, deciphers all the high-level software routines in which the machine-code instructions are embedded. An easier task will be a series of genome projects for different species (like the *Arabidopsis* plant genome project, whose completion is announced on the day that I write). These would be quicker and easier than the HGP, not because the other genomes are smaller or simpler than ours, but because the collective expertise of scientists increases cumulatively and rapidly with experience.

There is a frustrating aspect of this cumulative improvement. Given the rate of technological advance, with hindsight, when we started the Human Genome Project it wasn't worth starting. It would have been better to do nothing until the last two years and start then! Indeed, that is pretty much what the rival firm of Dr Craig Venter did. The fallacy in the 'never bother to start' maxim is that later technologies cannot get into a position to 'overtake' without the experience gained in developing the earlier ones.*

The HGP implicitly plays down the differences between individuals. But, with the intriguing exception of identical twins, everybody's genome is unique, and you might wonder *whose* genome is being sequenced in the HGP. Has some dignitary been singled out for the honour, is it a random person pulled off the street, or even an anonymous clone of cells in a tissue culture lab? It makes a difference. I have brown eyes while you have blue. I can't curl my tongue into a tube, whereas it's 50/50 that you can. Which version of the tongue-curling gene makes it into the published Human Genome? Which is the canonical eye colour? The answer is that, for the few 'letters' of the DNA text that vary, the canonical genome is the majority 'vote' among a sample of people carefully chosen to give a good spread of human diversity. But the diversity itself is expunged from the record.

By contrast the Human Genome Diversity Project (HGDP), now under way, builds on the foundation of the HGP but focuses on those relatively few nucleotide sites that *vary* from person to person, and from group to group. Incidentally, a surprisingly small proportion of that variance consists of between-race variance, a fact that has sadly failed to reassure spokesmen for various ethnic groups, especially in America. They have dreamed up influential political objections to the project which they see as exploitative and tarred with the brush of eugenics.

The medical benefits of studying human variation could be immense. Hitherto, almost all medical prescribing has assumed that patients are pretty much the same, and that every disease has an optimal recommended cure. Doctors of tomorrow will be more like vets in this respect. Doctors have only one species of patient,

*I have discussed the implications of the rapid growth of our understanding of genetics in more detail in 'Son of Moore's Law' (see pp. 127–36).

but in future they will subdivide that species by genotype, as a vet subdivides his patients by species. For the special needs of blood transfusions, doctors already recognize a few genetic typings (OAB, Rh) etc. In the future, every patient's personal record will include the results of numerous genetic tests: not their entire genome (that will be too expensive for the foreseeable future) but, as the century goes on, an increasing sampling of the variable regions of the genome, and far more than the present 'blood group' typings. The point is that for some diseases there may be as many different optimal treatments as there are different geno-types at a locus – more even, because genetic loci may *interact* to affect susceptibility to disease.

Another important use of the genetics of human diversity is forensic. Precisely because DNA is digital like computer bytes, genetic fingerprinting is potentially many many orders of magni-tude more accurate and reliable than any other means of individual identification, *including* direct facial recognition (despite the un-shakeable gut feeling of jurors that eyewitness identification trumps everything). Moreover, identity can be established from a tiny trace of blood, sweat or tears (or spit, semen or hairs).

DNA evidence is widely regarded as controversial, and I need to say a little about why. Firstly, human error can obviously vitiate the accuracy of the method. But that is true of all evidence. Courts are already accustomed to taking precautions to avoid the muddling up of specimens, and such precautions now become even more important. DNA fingerprinting can establish, almost infinitely far beyond all reasonable doubt, whether a smear of blood came from a particular individual. But obviously you must test the right smear.

Secondly, astronomical though the odds against mistaken identity by DNA fingerprinting theoretically are, it is possible for geneticists and statisticians to come up with what seem like widely different estimates of the precise odds. I quote from *Unweaving the Rainbow*[19] (Chapter 5, which is devoted to explain DNA fingerprinting in lay terms).

Lawyers are accustomed to pouncing when expert witnesses seem to disagree. If two geneticists are summoned to the stand and are asked to estimate the probability of a misidentification with DNA evidence, the first may say 1,000,000 to one while the second may say only 100,000 to one. Pounce. 'Aha! AHA! The experts disagree! Ladies and gentlemen of the jury, what confidence can we place in a scientific method, if the experts themselves can't get within a factor of ten of one another? Obviously the only thing to do is throw the entire evidence out, lock, stock and barrel.'

But ... any disagreement ... is only over whether the odds against a wrongful identification are hyper-mega-astronomical, or just plain astronomical. The odds cannot normally be lower than thousands to one, and they may well be up in the billions. Even on the most conservative estimate, the odds against wrongful identification are hugely greater than they are in an ordinary identity parade. 'M'lud, an identity parade of only 20 men is grossly unfair on my client. I demand a line-up of at least a million men!'

The idea of a nationwide database, in which all citizens' DNA fingerprints would be held, is now being discussed (only a sample of genes, of course: doing the whole genome would be overkill, far too expensive). I don't see this as a sinister, Big Brotherish idea (and I have written to my doctor volunteering to be a guinea pig in the pilot study of 500,000 now being prepared). But there are potential problems, of a civil liberties character. If your house is burgled, the police will routinely look for (traditional, old-fashioned) fingerprints of the burglar. They need to fingerprint the householder's family too, for elimination purposes, and most people are happy to oblige. Obviously the same principle will apply to DNA fingerprinting, but many people would want to stop well short of a nationwide database. Presumably they would also object to a nationwide database of conventional, old-fashioned finger-prints, but perhaps that is not a practical issue because it would take too long to search through it for a match. DNA fingerprinting doesn't suffer from this difficulty. Computer searches of huge DNA databases could be accomplished swiftly.

What, then, are the civil liberties problems? Surely, those with nothing to hide will have nothing to fear? Perhaps not, but some people do have legitimate reasons to hide information, not from the law but from each other. A surprisingly large number of people, of all ages, are genetically unrelated to the man they think is their father. To put it mildly, it is not clear that to disillusion them, with conclusive DNA evidence, would increase the sum of human happiness. If a national DNA database were in place, it might be hard to control unauthorized access to it. If a tabloid newspaper were to discover that the official heir to a Dukedom was actually sired by the gamekeeper, the consternation in the College of Heralds might be mildly amusing. But in the population at large it doesn't take much to imagine the family recriminations and sheer private misery that could flow from freely available information of true paternity. Nevertheless, the existence of a national DNA database wouldn't alter the situation much. It is already perfectly feasible for a jealous husband, say, to take a saliva or blood sample from one of his supposed children and compare it with his own, in order to confirm his suspicion that he is not the real father. What the national database could add is a swift computer search to find out who, out of all the males in the entire country, *is*!

More generally, the study of human diversity is one of very few areas where a good (though in my opinion not overwhelming) case can be made against the pure disinterested search for knowledge: one of very few areas where we might actually be better off ignorant. It is possible that, by the end of the twenty-first century, doctors will be able accurately to predict the manner and time of death of everybody, from the day they are conceived. At present this kind of deterministic prognostication can be achieved only for possessors of genes such as Huntington's Disease.* For the rest of us, all that is possible is the vague statistical forecast of the life insurance actuary, based on our smoking and drinking habits, and a quick listen through a stethoscope. The whole life insurance business depends upon such forecasts being vague and statistical. Those who die old subsidize (the heirs of) those who

---

*The folk singer Woody Guthrie died of Huntington's Disease, a horrible disease that waits till early middle age before killing you. It's a dominant gene, so each of Woody's children knows that he has an exactly 50 per cent chance of suffering the same horrible fate. Some people, given these odds, prefer not to be tested. They'd rather not know until they have to. IVF doctors can now push the test back to the newly fertilized zygote, and choose to implant only those that lack the fatal gene. This is obviously a huge boon, but it is attacked by ignorant lobbies fearful of 'scientists playing God'.

die young. If the day comes when deterministic forecasting (along Huntington's Disease lines) becomes universal, life insurance as we know it will collapse. That problem is soluble (presumably by universal compulsory life insurance with no individual medical risk assessment). What will be less easy to solve is the angst which will hang over everyone's psychology. As things are now, we all know we are going to die, but most of us don't know when, so it doesn't feel like a death *sentence*. That may change, and society should be prepared for difficulties as people struggle to adjust their psychologies to it.

Ethics

I have already touched on some ethical issues. Science has no methods for deciding what is ethical. That is a matter for individuals and for society. But science can clarify the questions being asked, and can clear up obfuscating misunderstandings. This usually amounts to the useful 'You cannot have it both ways' style of arguing. I'll give five examples, before turning to a more unusual interpretation of the phrase 'science and ethics'.

Science cannot tell you whether abortion is wrong, but it can point out that the (embryological) continuum that seamlessly joins a non-sentient foetus to a sentient adult is analogous to the (evolutionary) continuum that joins humans to other species. If the embryological continuum appears to be more seamless, this is only because the evolutionary continuum is divided by the accident of extinction. Fundamental principles of ethics should not depend on the accidental contingencies of extinction.* To repeat, science cannot tell you whether abortion is murder, but it can warn you that you may be being inconsistent if you think abortion is murder but killing chimpanzees is not. You cannot have it both ways.

Science cannot tell you whether it is wrong to clone a whole human being. But it can tell you that a Dolly-style clone is just an identical twin, though of a different age. It can tell you that, if you want to object to cloning humans, you must not appeal to

*See 'Gaps in the Mind' (pp. 23–30) for a fuller discussion.

arguments such as 'The clone wouldn't be a full person' or 'The clone wouldn't have a soul'. Science cannot tell you whether anybody has a soul, but it can tell you that, if ordinary identical twins have souls, so do Dolly-style clones.* You cannot have it both ways.

Science cannot tell you whether stem cell cloning for 'spare parts' is wrong. But it can challenge you to explain how stem cell cloning differs morally from something that has long been accepted: tissue culture. Tissue culture has been a mainstay of cancer research for decades. The famous HeLa cell line, which originated in the late **He**nrietta **La**cks in 1951, is now being grown in labs all over the world. A typical lab, at the University of California, grows 48 litres of HeLa cells per day, as a routine service to researchers in the university. The total daily worldwide production of HeLa cells must be measured in tons – all a gigantic clone of Henrietta Lacks. In the half century since this mass production began, nobody seems to have objected to it. Those who agitate to stop stem cell research today have to explain why they do not object to the mass cultivation of HeLa cells. You cannot have it both ways.

Science cannot tell you whether it is right to kill 'Mary' to save her conjoined twin 'Jodie' (or whether both twins should be allowed to die).† But science can tell you that a placenta is a true clone of the baby it nourishes. You could legitimately 'spin' the story of any placenta as a 'twin' of the baby that it nourishes, to be discarded when its role is completed. Admittedly, nobody is tempted to call their placenta Mary, but one might equally question the emotional wisdom of bestowing such a name on a Siamese twin with no heart or lungs, and only a primitive brain.

*See 'Dolly and the Cloth Heads' (pp. 180–3).
†These were widely publicized pseudonyms given to a pair of conjoined 'Siamese' twins who came to Britain for medical treatment around this time. The authorities wanted, against the parents' wishes, to separate the twins, in a mammoth operation which could have given Jodie (some sort of) life but would certainly result in Mary's death. Without the operation both twins would die, because Mary, who lacked most vital organs including a functioning brain, subsisted parasitically on Jodie. Many liberal people thought it right to over-rule the parents' religiously-based reluctance to 'kill' Mary to save Jodie. I thought the parents were right to reject the operation, although for the wrong reasons, and that in any case their wishes should have been respected because it was they whose lives were likely to be profoundly affected by the demands of the severely handicapped surviving twin.

And if anybody wishes to invoke 'slippery slopes' and 'thin ends of wedges' here, let them think on the following.

In 1998, a television gastronome served on screen a new gourmet dish: human placenta. He

> flash-fried strips of the placenta with shallots and blended two thirds into a puree. The rest was flambéed in brandy, and then sage and lime juice were added. The family of the baby concerned ate it, with twenty of their friends. The father thought it so delicious that he had fourteen helpings.

The whole thing was presented in the papers as a bit of a lark. Yet those who worry about slippery slopes need to ask themselves why that television dinner should not be called cannibalism. Cannibalism is one of our oldest and deepest taboos, and a devotee of the 'slippery slope' or 'thin end of the wedge' style of argument might do well to worry at the slightest breach of that taboo. I suspect that, if the television executives had known enough science to understand that a placenta *is* a true clone of a baby, the dinner would never have gone ahead, especially at the height of the Dolly-inspired cloning controversy. You cannot have it both ways.

I want to conclude with a rather idiosyncratic approach to the matter of science and ethics: ethical treatment of scientific truth itself. I want to suggest that objective truth sometimes needs the same kind of protection as the libel laws now give to individuals. Or at least to suggest that the Trades Descriptions Act might be more imaginatively invoked. I'll say a little about this first, in the light of Prince Charles's recent plea for public money to do research into 'alternative medicine'.

If a pharmaceutical company advertises its pills as curing headaches, it must be able to demonstrate, in double-blind controlled trials, that its pills do indeed cure headaches. Double-blind means, of course, that neither the patients, nor the testers, know until afterwards which patients received the dose, and which the placebo control. If the pills cannot pass this test – if numerous strenuous efforts fail to distinguish them from a neutral placebo – I presume the company might be in danger of prosecution under the Trades Descriptions Act.

Homeopathic remedies are big business, they are advertised as efficacious in various ways, yet they have never been demonstrated to have any effect at all. Personal testimony is ubiquitous, but it is useless evidence because of the notorious power of the placebo effect. This is exactly why 'orthodox' medicines are obliged to prove themselves in double-blind trials.*

I do not want to imply that all so-called 'alternative medicines' are as useless as homeopathy. For all I know, some of them may work. But they must be *demonstrated* to work, by double-blind placebo-control trials or some equivalent experimental design. And if they pass that test, there is then no longer any reason to call them 'alternative'. Mainstream medicine would simply adopt them. As the late distinguished journalist John Diamond wrote movingly (like many patients dying of cancer, he had false hopes cruelly raised by a succession of plausible quacks) in The *Independent*:

> There is really no such thing as alternative medicine, just medicine that works and medicine that doesn't ... There isn't an 'alternative' physiology or anatomy or nervous system any more than there's an alternative map of London which lets you get to Battersea from Chelsea without crossing the Thames.

But I began this final section in more radical terms. I wanted to extend the concept of libel to include lies that may not damage particular people but damage truth itself. Some twenty years ago, long before Dolly showed it was plausible, a book was published claiming, in great detail, that a rich man in South America had had himself cloned, by a scientist code-named Darwin. As a work of science fiction it would have been unexceptionable, but it was sold as sober fact. The author and publishers were sued, by Dr Derek Bromhall, who claimed that his reputation as a scientist was damaged by his being quoted in the book. My point is that whatever damage may or may not have been done to Dr Bromhall, far more important was the damage done to scientific truth itself.

That book has faded from memory and I bring it up only as an example. Obviously I want to generalize the principle to all deliberate falsifications, misrepresentations, of scientific truth.

*Homeopathy has special problems with double-blind control testing. I discuss this in my Foreword to John Diamond's *Snake Oil* (see pp. 211–15).

Why should a Derek Bromhall have to prove himself personally damaged, before we can prosecute a book which wantonly publishes lies about the universe? As will be obvious I'm no lawyer but, if I was, rather than constantly feel the need to drag things down to the question of whether particular humans have been damaged, I think I would like to stand up and defend truth itself. No doubt I shall be told – and convinced – that a court of law is not the right place for this. But in the wider world, if I am asked for a single phrase to characterize my role as Professor of Public Understanding of Science, I think I would choose Advocate for Disinterested Truth.

1.5

Trial By Jury[20]

Trial by jury must be one of the most conspicuously bad good ideas anyone ever had. Its devisers can hardly be blamed. They lived before the principles of statistical sampling and experimental design had been worked out. They weren't scientists. Let me explain using an analogy. And if, at the end, somebody objects to my argument on the grounds that humans aren't herring gulls, I'll have failed to get my point across.

Adult herring gulls have a bright yellow bill with a conspicuous red spot near the tip. Their babies peck at the red spot, which induces the parents to regurgitate food for them. Niko Tinbergen, Nobel Prize-winning zoologist and my old maestro at Oxford, offered naive young chicks a range of cardboard dummy gull heads varying in bill and spot colour, and shape. For each colour, shape or combination, Tinbergen measured the preferences of the baby chicks by counting their pecks in a standard time. The idea was to discover whether naive gull chicks are born with a built-in preference for long yellow things with red spots. If so, this would suggest that genes equip the young birds with detailed prior knowledge of the world in which they are about to hatch – a world in which food comes out of adult herring gull beaks.

Never mind the reason for the research, and never mind the conclusions. Consider, instead, the methods you must use, and the pitfalls you must avoid, if you want to get a correct result in any such experiment. These turn out to be general principles which apply to human juries as strongly as to gull chicks.

First, you obviously must test more than one chick. It could be that some chicks are red-biased, others blue-biased, with no tendency for herring gull chicks in general to share the same favourite colour. So, by picking out a single chick, you are measuring nothing more than individual bias.

So, we must test more than one chick. How many? Is two enough? No, nor is three, and now we must start to think statistically. To make it simple, suppose that in a particular experiment we are comparing only red spots versus blue spots, both on a yellow background, and always presented simultaneously. If we test just two chicks separately, suppose the first chick chooses red. It had a 50 per cent chance of doing so, at random. Now the second chick also happens to choose red. Again, the odds were 50 per cent that it would do so at random, even if it were colour-blind. There's a 50 per cent chance that two randomly choosing chicks will agree (half of the four possibilities: red red, red blue, blue red, blue blue). Three chicks aren't enough either. If you write down all the possibilities, you'll find that there's a 25 per cent chance of a unanimous verdict, by luck alone. Twenty-five per cent, as the odds of reaching a conclusion for the wrong reason, is unacceptably large.

How about twelve good chicks and true? Now you're talking. If twelve chicks are independently offered a choice between two alternatives, the odds that they will all reach the same verdict by chance alone are satisfyingly low, only one in 1024.

But now suppose that, instead of testing our twelve chicks independently, we test them as a group. We take a maelstrom of twelve cheeping chicks and lower into their midst a red spotted dummy and a blue spotted dummy, each fitted with an electrical device for automatically tallying pecks. And suppose that the collective of chicks registers 532 pecks at red and zero at blue. Does this massive disparity show that those twelve chicks prefer red? Absolutely not. The pecks are not independent data. Chicks could have a strong tendency to imitate one another (as well as imitate themselves in lock-on effects). If one chick just happened to peck at red first, others might copy him and the whole company of chicks join in a frenzy of imitative pecking. As a matter of fact this is precisely what domestic chicken chicks do, and gull chicks are very likely the same. Even if not, the principle remains that the data are not independent and the experiment is therefore invalid. The twelve chicks are strictly equivalent to a single chick, and their summed pecks, however numerous, might as well be only a single peck: they amount to only a single independent result.

Turning to courts of law, why are twelve jurors preferred to a single judge? Not because they are wiser, more knowledgeable or more practised in the arts of reasoning. Certainly not, and with a vengeance. Think of the astronomical damages awarded by juries in footling libel cases. Think how juries bring out the worst in histrionic, gallery-playing lawyers. Twelve jurors are preferred to one judge only because they are more numerous. Letting a single judge decide a verdict would be like letting a single chick speak for the whole herring gull species. Twelve heads are better than one, because they represent twelve assessments of the evidence.

But for this argument to be valid, the twelve assessments really have to be independent. And of course they are not. Twelve men and women locked in a jury room are like our clutch of twelve gull chicks. Whether they actually imitate each other like chicks, they might. That is enough to invalidate the principle by which a jury might be preferred over a single judge.

In practice, as is well documented and as I remember from the three juries that it has been my misfortune to serve on, juries are massively swayed by one or two vocal individuals. There is also strong pressure to conform to a unanimous verdict, which further undermines the principle of independent data. Increasing the number of jurors doesn't help, or not much (and not at all, in strict principle). What you have to increase is the number of *independent* verdict-reaching units.

Oddly enough, the bizarre American system of televising trials opens up a real possibility of improving the jury system. By the end of trials such as those of Louise Woodward or O. J. Simpson, literally thousands of people around the country have attended to the evidence as assiduously as the official jury. A mass phone-in might produce a fairer verdict than a jury. But, unfortunately, journalistic discussion, radio talk-shows and ordinary gossip would violate the Principle of Independent Data and we'd be back where we started. The broadcasting of trials, in any case, has horrible consequences. In the wake of Louise Woodward's trial, the Internet seethed with ill-spelled and ungrammatical viciousness, the cheque-book journalists were queuing up, and the unfortunate judge presiding had to change his telephone number and employ a bodyguard.

So, how can we improve the system? Should twelve jurors be locked in twelve isolation chambers and their opinions separately polled so that they constitute genuinely independent data? If it is objected that some would be too stupid or inarticulate to reach a verdict on their own, we are left wondering why such individuals are allowed on a jury at all. Perhaps there is something to be said for the collective wisdom that emerges when a group of people thrash out a topic together, round a table. But this still leaves the principle of independent data unsatisfied.

Should all cases be tried by two separate juries? Or three? Or twelve? Too expensive, at least if each jury has twelve members. Two juries of six members, or three juries of four members, would probably be an improvement over the present system. But isn't there some way of testing the relative merits of such alternative options, or of comparing the merits of trial by jury versus trial by judge?

Yes, there is. I'll call it the Two Verdicts Concordance Test. It is based on the principle that, if a decision is valid, two independent shots at making it should yield the same result. Just for purposes of the test, we run to the expense of having two juries, listening to the same case and forbidden to talk to members of the other jury. At the end, we lock the two juries in two separate jury rooms and see if they reach the same verdict. If they don't, neither verdict has been proved beyond reasonable doubt, and this would cast reasonable doubt on the jury system itself.

To make the experimental comparison with Trial by Judge, we need two experienced judges to listen to the same case, and require them too to reach their separate verdicts without talking to each other. Whichever system, Trial by Jury or Trial by Judge, yields the higher score of agreements over a number of trials is the better system and might, if its Concordance Score is high, even be accredited for future use with some confidence.

Would you bet on two independent juries reaching the same verdict in the Louise Woodward case? Could you imagine even *one* other jury reaching the same verdict in the O. J. Simpson case? Two judges, on the other hand, seem to me rather likely to score well on the concordance test. And should I be charged with a serious crime, here's how I want to be tried. If I know myself to be guilty, I'll go with the loose cannon of a jury, the more ignorant, prejudiced and capricious the better. But if I am innocent, and the ideal of multiple independent decision-takers is unavailable, please give me a judge.

1.6

Crystalline Truth
and Crystal Balls[21]

A celebrated film star 'places four quartz crystal clusters in the four corners of her bathtub every time she takes a bath'. This doubtless has some mystic connection with the following recipe for meditation.

Each of the four quartz crystals in the meditation room should be 'programmed' to project gentle, loving, relaxing, crystalline energy towards all those present within the Meditation group. The quartz crystals will then generate a field of positive crystalline energy surrounding everyone in the room.

Language like this is a con-trick. It sounds 'scientific' enough to bamboozle the innocent. 'Programming' is what you do to computers. The word means nothing when applied to crystals. 'Energy' and 'field' are carefully defined notions in physics. There is no such thing as 'loving' or 'crystalline' energy, whether positive or no.*

New Age lore also advises placing a quartz crystal in your water jug. 'You will soon appreciate the sparkling purity of your crystal water.' See how the trick works. Somebody with no understanding of the real world could make a kind of 'poetic' association with 'crystal clear' water. But that is no more sensible than trying to read by the light of a ('bright as a') button. Or putting ('hard as') nails under your pillow to assist an erection.

Try the following experiment when you next suffer from 'flu': hold your personal quartz crystal and visualize yellow light radiating through it. Then place your crystal in a jug of water and drink this water the next day; one cup of water at two-hourly intervals. You will be amazed at the result!

*And, by the way, the next time you visit an 'alternative' therapist who claims to be 'balancing your energy fields', challenge them to say what they mean. The answer will be absolutely nothing.

Drinking water at two-hourly intervals is a good idea anyway, when you have flu. Putting a quartz crystal in it will have no additional effect. In particular, no amount of 'visualizing' of coloured light will change the composition of either the crystal or the water.

Pseudoscientific drivel like this is a disturbingly prominent part of the culture of our age. I have limited my examples to crystals because I had to draw a line somewhere. But 'star signs' would have done just as well. Or 'angels', 'channelling', 'telepathy', 'quantum healing', 'homeopathy', 'map-dowsing'. There is no obvious limit to human gullibility. We are docile credulity-cows, eager victims of quacks and charlatans who milk us and grow fat. There is a rich living to be made by anyone prepared to prostitute the language – and the wonder – of science.

But isn't it all – crystal ball gazing, star signs, birth stones, ley-lines and the rest – just a bit of harmless fun? If people want to believe in garbage like astrology, or crystal healing, why not let them? But it's so sad to think about all that they are *missing*. There is so much wonder in real science. The universe is mysterious enough to need no help from warlocks, shamans and 'psychic' tricksters. These are at best a soul-sapping distraction. At worst they are dangerous profiteers.

The real world, properly understood in the scientific way, is deeply beautiful and unfailingly interesting. It's worth putting in some honest effort to understand it properly, undistracted by false wonder and prostituted pseudoscience. For illustration, we need look no further than crystals themselves.

In a crystal such as quartz or diamond the atoms are arranged in a precisely repeating pattern. The atoms in a diamond – all identical carbon atoms – are arrayed like soldiers on parade except that the precision of their dressing far outsmarts the best-drilled guards regiment, and the atomic soldiers outnumber all the people that have ever lived or ever will. Imagine yourself shrunk to become one of the carbon atoms in the heart of a diamond crystal. You are one of the soldiers in a gigantic parade, but it'll seem a little odd because the files are arrayed in three dimensions. Perhaps a prodigious school of fish is a better image.

Each fish in the school is one carbon atom. Think of them

hovering in space, keeping their distance from each other and holding their precise angles, by means of forces that you can't see but which scientists fully understand. But if this is a fish school, it is one that – to scale – would fill the Pacific Ocean. In any decent-sized diamond, you are likely to be looking along arrays of atoms numbering hundreds of millions in any one straight line.

Carbon atoms can take up other crystal lattice formations. To revert to the military analogy, they can adopt alternative drill conventions. Graphite (the 'lead' in pencils) is also carbon, but it's obviously nothing like diamond. In graphite, the atoms form sheets of hexagons, like chicken wire. Each sheet is loosely bonded to those above and below it, and when impurities are present the sheets slide easily against each other, which is why graphite is a good lubricant. Diamond is very much not a lubricant. Its legendary hardness abrades the toughest materials. The atoms in soft graphite and hard diamond are identical. If you could persuade the atoms in graphite crystals to adopt the drill rules of diamond crystals, you'd be rich. It can be done, but you need colossal pressures and high temperatures, presumably the conditions that naturally manufacture diamonds, deep in the earth.

If hexagons make a sheet of flat graphite, you can imagine that interspersing some pentagons among the hexagons could make the sheet buckle into a curve. Place exactly 12 pentagons strategically among 20 hexagons and the curve bends round into a complete sphere. Geometers call it a truncated icosahedron. This is exactly the pattern of the sewing seams on a football. The football is, therefore, theoretically a pattern into which carbon atoms might spontaneously fall.

Mirabile dictu, exactly this pattern has been discovered among carbon atoms. The team responsible, including Sir Harry Kroto of Sussex University, won the 1996 Nobel Prize for Chemistry. Called Buckminsterfullerene, it is an elegant sphere of 60 carbon atoms, linked up as 20 hexagons interspersed with 12 pentagons. The name honours the visionary American architect Buckminster Fuller (whom I was privileged to meet when he was a very old man*) and the spheres are affectionately known as buckyballs.

*He was billed to give us a short lecture but, unscripted, he held us spellbound for three hours.

They can combine together to make larger crystals. Like graphite sheets, buckyballs make good lubricants, probably because of their spherical shape: they presumably work like tiny ball bearings.

Since the buckyball's discovery, chemists have realized that it is just a special case of a large family of 'buckytubes' and other 'fullerenes'. Carbon atoms can theoretically join up to form an Aladdin's cave of fascinating crystalline forms – another aspect of the unique property that qualifies carbon to be the fundamental element of life.

Not every atom has carbon's talent for joining copies of itself. Other crystals contain more than one kind of 'soldier', alternating in some elegant pattern. In quartz crystals it is silicon and oxygen instead of carbon; in common salt it is electrically charged atoms of sodium and chlorine. Crystals naturally break along lines that betray the underlying regimental drill pattern. That is why salt crystals are square, why the honeycomb columns of the Giant's Causeway stand as they do, and why diamond crystals are, well, diamond-shaped.

All crystals 'self-assemble' under locally acting rules. Their component 'soldiers', floating in free solution in water, spontaneously plug themselves into 'gaps' on the surface of the existing crystal, where they exactly fit. So a crystal may grow in solution from a tiny 'seed' – perhaps an impurity like the sand grain at the heart of a pearl. There is no grand design of buckyballs, quartz crystals, diamonds or anything else. This principle of self-assembly runs right through living structure, too. DNA itself (the genetic molecule, the molecule at the centre of all life) can be regarded as a long, spiral crystal in which one half of the double helix self-assembles on a template provided by the other. Viruses self-assemble like elaborately complex crystal-clusters. The head of the T4 bacteriophage (a virus that infects bacteria) actually looks like a single crystal.

Go into any museum and look at the collection of minerals. Even go into a New Age shop and look at the crystals on display, along with all the other apparatus of mumbo-jumbo and kitsch con-trickery. The crystals won't respond to your attempts to 'program' them for meditation, or 'dedicate' them with warm, loving thoughts. They won't cure you of anything, or fill the room with 'inner peace' or 'psychic energy'. But many of them are very beautiful, and it surely only adds to the beauty when we understand that the shapes of the crystals, the angles of their facets, the rainbow colours that flash from inside them, all have a precise explanation which lies deep in the patterns of atomic lattice-work.

Crystals don't vibrate with mystical, loving energy. But they do, in a much stricter and more interesting sense, vibrate. Some crystals have an electric charge across them, which changes when you physically deform the crystal. This 'piezo-electric' effect, discovered in 1880 by the Curie brothers (Marie's husband and his brother), is used in the styluses of record players (the 'deforming' is done by the groove of the turning record) and in some microphones (the deforming is done by sound waves in the air). The piezo effect works in reverse. When a suitable crystal is placed in an electric field it deforms itself rhythmically. Often the timing of this oscillation is extremely accurate. It serves as the equivalent of the pendulum or balance wheel in a quartz watch.

Let me tell you one last thing about crystals, and it may be the most fascinating of all. The military metaphor makes us think of each soldier as a metre or two from his neighbours. But actually almost all the interior of a crystal is empty space. My head is 18 centimetres in diameter. To keep to scale, my nearest neighbours in the crystalline parade would have to be standing more than a kilometre away. No wonder the tiny particles called neutrinos (even smaller than electrons) pass right through the earth and come out the other side as if it wasn't there (on average one passes through you every second).

But if solid things are mostly empty space, why don't we see them as empty space? Why does a diamond feel hard and solid instead of crumbly and full of holes? The answer lies in our own evolution. Our sense organs, like all our bits, have been shaped by Darwinian natural selection over countless generations. You might think that our sense organs would be shaped to give us a 'true' picture of the world as it 'really' is. It is safer to assume that they have been shaped to give us a *useful* picture of the world, to help us to survive. In a way, what sense organs do is assist our brains to construct a useful model of the world, and it is this model that we move around in. It is a kind of 'virtual reality' simulation of the real world. Neutrinos can pass straight through a rock but we can't. If we try to, we hurt ourselves. When constructing its simulation of rock, the brain therefore represents it as hard and solid. It's almost as though our sense organs are telling us: 'You can't get through objects of this kind.' That's what 'solid' means. That's why we perceive them as 'solid'.

In the same way we find much of the universe, as science discovers it, difficult to understand. Einstein's relativity, quantum uncertainty, black holes, the big bang, the expanding universe, the vast slow movement of geological time – all these are hard to grasp. No wonder science frightens some people. But science can even explain why these things are hard to understand, and why the effort frightens us. We are jumped-up apes, and our brains were only designed to understand the mundane details of how to survive in the stone-age African savannah.

These are deep matters, and a short article is not the place to go into them. I shall have succeeded if I have persuaded you that a scientific approach to crystals is more illuminating, more uplifting, and also stranger, than anything imagined in the wildest dreams of New Age gurus or paranormal preachers. The blunt truth is that the dreams and visions of gurus and preachers are not nearly wild enough. By scientific standards, that is.

Postmodernism Disrobed[22]

Review of *Intellectual Impostures*
by Alan Sokal and Jean Bricmont

Suppose you are an intellectual impostor with nothing to say, but with strong ambitions to succeed in academic life, collect a coterie of reverent disciples and have students around the world anoint your pages with respectful yellow highlighter. What kind of literary style would you cultivate? Not a lucid one, surely, for clarity would expose your lack of content. The chances are that you would produce something like the following:

> We can clearly see that there is no bi-univocal correspondence between linear signifying links or archi-writing, depending on the author, and this multireferential, multi-dimensional machinic catalysis. The symmetry of scale, the transversality, the pathic non-discursive character of their expansion: all these dimensions remove us from the logic of the excluded middle and reinforce us in our dismissal of the ontological binarism we criticised previously.

This is a quotation from the psychoanalyst Félix Guattari, one of many fashionable French 'intellectuals' outed by Alan Sokal and Jean Bricmont in their splendid book *Intellectual Impostures*, which caused a sensation when published in French last year, and which is now released in a completely rewritten and revised English edition. Guattari goes on indefinitely in this vein and offers, in the opinion of Sokal and Bricmont, 'the most brilliant mélange of scientific, pseudo-scientific and philosophical jargon that we have ever encountered'. Guattari's close collaborator, the late Gilles Deleuze had a similar talent for writing:

> In the first place, singularities-events correspond to heterogeneous series which are organized into a system which is neither stable nor unstable, but rather 'metastable', endowed with a potential energy

wherein the differences between series are distributed ... In the second place, singularities possess a process of auto-unification, always mobile and displaced to the extent that a paradoxical element traverses the series and makes them resonate, enveloping the corresponding singular points in a single aleatory point and all the emissions, all dice throws, in a single cast.

It calls to mind Peter Medawar's earlier characterization of a certain type of French intellectual style (note, in passing, the contrast offered by Medawar's own elegant and clear prose):

Style has become an object of first importance, and what a style it is! For me it has a prancing, high-stepping quality, full of self-importance, elevated indeed, but in the balletic manner, and stopping from time to time in studied attitudes, as if awaiting an outburst of applause. It has had a deplorable influence on the quality of modern thought ...

Returning to attack the same targets from another angle, Medawar says:

I could quote evidence of the beginnings of a whispering campaign against the virtues of clarity. A writer on structuralism in the *Times Literary Supplement* has suggested that thoughts which are confused and tortuous by reason of their profundity are most appropriately expressed in prose that is deliberately unclear. What a preposterously silly idea! I am reminded of an air-raid warden in wartime Oxford who, when bright moonlight seemed to be defeating the spirit of the blackout, exhorted us to wear dark glasses. He, however, was being funny on purpose.

This is from Medawar's 1968 Lecture on 'Science and Literature', reprinted in *Pluto's Republic*[23]. Since Medawar's time, the whispering campaign has raised its voice.

Deleuze and Guattari have written and collaborated on books described by the celebrated Michel Foucault as 'among the greatest of the great ... Some day, perhaps, the century will be Deleuzian.' Sokal and Bricmont, however, remark that

These texts contain a handful of intelligible sentences – sometimes banal, sometimes erroneous – and we have commented on some of them in the footnotes. For the rest, we leave it to the reader to judge.

But it's tough on the reader. No doubt there exist thoughts so profound that most of us will not understand the language in which they are expressed. And no doubt there is also language designed to be unintelligible in order to conceal an absence of honest thought. But how are we to tell the difference? What if it really takes an expert eye to detect whether the emperor has clothes? In particular, how shall we know whether the modish French 'philosophy', whose disciples and exponents have all but taken over large sections of American academic life, is genuinely profound or the vacuous rhetoric of mountebanks and charlatans?

Sokal and Bricmont are professors of physics at, respectively, New York University and the University of Louvain. They have limited their critique to those books that have ventured to invoke concepts from physics and mathematics. Here they know what they are talking about, and their verdict is unequivocal: on Lacan, for example, whose name is revered by many in humanities departments throughout American and British universities, no doubt partly because he simulates a profound understanding of mathematics:

> ... although Lacan uses quite a few key words from the mathematical theory of compactness, he mixes them up arbitrarily and without the slightest regard for their meaning. His 'definition' of compactness is not just false: it is gibberish.

They go on to quote the following remarkable piece of reasoning by Lacan:

> Thus, by calculating that signification according to the algebraic method used here, namely:
>
> $$\frac{S \text{ (signifier)}}{s \text{ (signified)}} = s \text{ (the statement)}$$
>
> With S=(-1), produces: $s = \sqrt{-1}$

You don't have to be a mathematician to see that this is ridiculous. It recalls the Aldous Huxley character who proved the existence of God by dividing zero into a number, thereby deriving the infinite. In a further piece of reasoning which is entirely typical of the *genre*, Lacan goes on to conclude that the erectile organ

> ... is equivalent to the √-1 of the signification produced above, of the *jouissance* that it restores by the coefficient of its statement to the function of lack of signifier (-1).

We do not need the mathematical expertise of Sokal and Bricmont to assure us that the author of this stuff is a fake. Perhaps he is genuine when he speaks of non-scientific subjects? But a philosopher who is caught equating the erectile organ to the square root of minus one has, for my money, blown his credentials when it comes to things that I *don't* know anything about.

The feminist 'philosopher' Luce Irigaray is another who is given whole chapter treatment by Sokal and Bricmont. In a passage reminiscent of a notorious feminist description of Newton's *Principia* (a 'rape manual'), Irigaray argues that $E=mc^2$ is a 'sexed equation'. Why? Because 'it *privileges* the speed of light over other speeds that are vitally necessary to us' (my emphasis of what I am rapidly coming to learn is an in-word). Just as typical of the school of thought under examination is Irigaray's thesis on fluid mechanics. Fluids, you see, have been unfairly neglected. 'Masculine physics' *privileges* rigid, solid things. Her American expositor Katherine Hayles made the mistake of re-expressing Irigaray's thoughts in (comparatively) clear language. For once, we get a reasonably unobstructed look at the emperor and, yes, he has no clothes:

> The privileging of solid over fluid mechanics, and indeed the inability of science to deal with turbulent flow at all, she attributes to the association of fluidity with femininity. Whereas men have sex organs that protrude and become rigid, women have openings that leak menstrual blood and vaginal fluids ... From this perspective it is no wonder that science has not been able to arrive at a successful model for turbulence. The problem of turbulent flow cannot be solved because the conceptions of fluids (and of women) have been formulated so as necessarily to leave unarticulated remainders.

You don't have to be a physicist to smell out the daffy absurdity of this kind of argument (the tone of it has become all too familiar), but it helps to have Sokal and Bricmont on hand to tell us the real reason why turbulent flow is a hard problem (the Navier-Stokes equations are difficult to solve).

In similar manner, Sokal and Bricmont expose Bruno Latour's confusion of relativity with relativism, Lyotard's 'postmodern science', and the widespread and predictable misuses of Gödel's Theorem, quantum theory and chaos theory. The renowned Jean Baudrillard is only one of many to find chaos theory a useful tool for bamboozling readers. Once again, Sokal and Bricmont help us by analysing the tricks being played. The following sentence, 'though constructed from scientific terminology, is meaningless from a scientific point of view':

> Perhaps history itself has to be regarded as a chaotic formation, in which acceleration puts an end to linearity and the turbulence created by acceleration deflects history definitively from its end, just as such turbulence distances effects from their causes.

I won't quote any more, for, as Sokal and Bricmont say, Baudrillard's text 'continues in a gradual crescendo of nonsense'. They again call attention to 'the high density of scientific and pseudoscientific terminology – inserted in sentences that are, as far as we can make out, devoid of meaning'. Their summing up of Baudrillard could stand for any of the authors criticized here, and lionized throughout America:

> In summary, one finds in Baudrillard's works a profusion of scientific terms, used with total disregard for their meaning and, above all, in a context where they are manifestly irrelevant. Whether or not one interprets them as metaphors, it is hard to see what role they could play, except to give an appearance of profundity to trite observations about sociology or history. Moreover, the scientific terminology is mixed up with a non-scientific vocabulary that is employed with equal sloppiness. When all is said and done, one wonders what would be left of Baudrillard's thought if the verbal veneer covering it were stripped away.

But don't the postmodernists claim only to be 'playing games'? Isn't it the whole point of their philosophy that anything goes, there is no absolute truth, anything written has the same status as anything else, no point of view is privileged? Given their own standards of relative truth, isn't it rather unfair to take them to task for fooling around with word-games, and playing little jokes on readers? Perhaps, but one is then left wondering why their writings are so stupefyingly boring. Shouldn't games at least be entertaining, not po-faced, solemn and pretentious? More tellingly, if they are only joking around, why do they react with such shrieks of dismay when somebody plays a joke at their expense? The genesis of *Intellectual Impostures* was a brilliant hoax perpetrated by Alan Sokal, and the stunning success of his *coup* was not greeted with the chuckles of delight that one might have hoped for after such a feat of deconstructive game playing. Apparently, when you've become the establishment, it ceases to be funny when somebody punctures the established bag of wind.

As is now rather well known, in 1996 Sokal submitted to the American journal *Social Text* a paper called 'Transgressing the Boundaries: towards a transformative hermeneutics of quantum gravity'. From start to finish the paper was nonsense. It was a carefully crafted parody of postmodern metatwaddle. Sokal was inspired to do this by Paul Gross and Norman Levitt's *Higher Superstition: the academic left and its quarrels with science*, an important book which deserves to become as well known in Britain as it already is in America. Hardly able to believe what he read in this book, Sokal followed up the references to postmodern literature, and found that Gross and Levitt did not exaggerate. He resolved to do something about it. In Gary Kamiya's words:

Anyone who has spent much time wading through the pious, obscurantist, jargon-filled cant that now passes for 'advanced' thought in the humanities knew it was bound to happen sooner or later: some clever academic, armed with the not-so-secret passwords ('hermeneutics', 'transgressive', 'Lacanian', 'hegemony', to name but a few) would write a completely bogus paper, submit it to an *au courant* journal, and have it accepted ... Sokal's piece uses all the right terms. It cites all the best people. It whacks sinners (white men, the 'real world'), applauds the virtuous (women, general metaphysical lunacy) ... And it is complete, unadulterated bullshit – a fact that somehow escaped the attention of the high-powered editors of *Social Text*, who must now be experiencing that queasy sensation that afflicted the Trojans the morning after they pulled that nice big gift horse into their city.

Sokal's paper must have seemed a gift to the editors because this was a *physicist* saying all the right-on things they wanted to hear, attacking the 'post-Enlightenment hegemony' and such uncool notions as the existence of the real world. They didn't know that Sokal had also crammed his paper with egregious scientific howlers, of a kind that any referee with an undergraduate degree in physics would instantly have detected. It was sent to no such referee. The editors, Andrew Ross and others, were satisfied that its ideology conformed to their own, and were perhaps flattered by references to their own works. This ignominious piece of editing rightly earned them the 1996 Ig Nobel Prize for literature.

Notwithstanding the egg all over their faces, and despite their feminist pretensions, these editors are dominant males in the academic lekking arena. Andrew Ross himself has the boorish, tenured confidence to say things like, 'I am glad to be rid of English Departments. I hate literature, for one thing, and English departments tend to be full of people who love literature'; and the yahooish complacency to begin a book on 'science studies' with these words: 'This book is dedicated to all of the science teachers I never had. It could only have been written without them.' He and his fellow 'cultural studies' and 'science studies' barons are not harmless eccentrics at third-rate state colleges. Many of them have tenured professorships at some of America's best universities. Men of this kind sit on appointment committees, wielding power over young academics who might secretly aspire to an *honest* academic career in literary studies or, say, anthropology. I know – because many of them have told me – that there are sincere scholars out there who would speak out if they dared, but who are intimidated into silence. To them, Alan Sokal will appear as a hero, and nobody with a sense of humour or a sense of justice will disagree. It helps, by the way, although it is strictly irrelevant, that his own left-wing credentials are impeccable.

In a detailed post-mortem of his famous hoax, submitted to *Social Text* but predictably rejected by them and published elsewhere, Sokal notes that, in addition to numerous half-truths, falsehoods and non sequiturs, his original article contained some 'syntactically correct sentences that have no meaning whatsoever'. He regrets that there were not more of the latter: 'I tried hard to produce them, but

I found that, save for rare bursts of inspiration, I just didn't have the knack.' If he were writing his parody today, he'd surely have been helped by a virtuoso piece of computer programming by Andrew Bulhak of Melbourne: the Postmodernism Generator. Every time you visit it at http://www.elsewhere.org/cgi-bin/postmodern/ it will spontaneously generate for you, using faultless grammatical principles, a spanking new postmodern discourse, never before seen. I have just been there, and it produced for me a 6000-word article called 'Capitalist theory and the subtextual paradigm of context' by 'David I. L. Werther and Rudolf du Garbandier of the Department of English, Cambridge University' (poetic justice there, for it was Cambridge who saw fit to give Jacques Derrida an honorary degree). Here's a typical sentence from this impressively erudite work:

> If one examines capitalist theory, one is faced with a choice: either reject neotextual materialism or conclude that society has objective value. If dialectic desituationism holds, we have to choose between Habermasian discourse and the subtextual paradigm of context. It could be said that the subject is contextualised into a textual nationalism that includes truth as a reality. In a sense, the premise of the subtextual paradigm of context states that reality comes from the collective unconscious.

Visit the Postmodernism Generator. It is a literally infinite source of randomly generated syntactically correct nonsense; distinguishable from the real thing only in being more fun to read. You could generate thousands of papers per day, each one unique and ready for publication, complete with numbered endnotes. Manuscripts should be submitted to the 'Editorial Collective' of *Social Text*, double-spaced and in triplicate.

As for the harder task of reclaiming humanities and social studies departments for genuine scholars, Sokal and Bricmont have joined Gross and Levitt in giving a friendly and sympathetic lead from the world of science. We must hope that it will be followed.

1.8

The Joy of Living Dangerously:
Sanderson of Oundle[24]

My life has lately been dominated by education. Home life over-shadowed by A-level* examination horrors, I escaped to London to address a conference of schoolteachers. On the train, in preparation for the inaugural 'Oundle Lecture' which I was nervously to give at my old school† the following week, I read H. G. Wells's biography of its famous Head: *The Story of a Great Schoolmaster: being a plain account of the life and ideas of Sanderson of Oundle.*[25] The book begins in terms which initially seemed a little over the top: 'I think him beyond question the greatest man I have ever known with any degree of intimacy.' But it led me on to read the official biography, *Sanderson of Oundle,*[26] written by a large, anonymous syndicate of his former pupils (Sanderson believed in cooperation instead of striving for individual recognition).

I now see what Wells meant. And I am sure that Frederick William Sanderson (1857–1922) would have been horrified to learn what I learned from the teachers I met at the London conference: about the stifling effects of exams, and the government obsession with measuring a school's performance by them. He would have been aghast at the anti-educational hoops that young people now have to jump through in order to get into university. He would have been openly contemptuous of the pussyfooting, lawyer-driven fastidiousness of 'Health and Safety', and the accountant-driven league tables that dominate modern education and actively

*Advanced-levels: school-leaving examinations, on which acceptance to British universities largely depends. A-levels notoriously traumatize teenagers, because so much hangs on the result. Schools vie with each other in nationally compiled tables of A-level performance, and ambitious schools have been known to discourage less able pupils from even trying, for fear of damaging the school's rank in the league table.
†Oundle School, in Northamptonshire in central England, founded 1556.

encourage schools to put their own interests before those of their pupils. Quoting Bertrand Russell, he disliked competition and 'possessiveness' as a motive for anything in education.

Sanderson of Oundle ended up second only to Arnold of Rugby in fame, but Sanderson was not born to the world of public schools. Today, he would, I dare say, have headed a large, mixed Comprehensive.* His humble origins, northern accent and lack of Holy Orders gave him a rough ride with the classical 'dominies' whom he found on arrival at the small and run-down Oundle of 1892. So rebarbative were his first five years, Sanderson actually wrote out his letter of resignation. Fortunately, he never sent it. By the time of his death thirty years later, Oundle's numbers had increased from 100 to 500, it had become the foremost school for science and engineering in the country, and he was loved and respected by generations of grateful pupils and colleagues. More important, Sanderson developed a philosophy of education that we should urgently heed today.

He was said to lack fluency as a public speaker, but his sermons in the School Chapel could achieve Churchillian heights:

Mighty men of science and mighty deeds. A Newton who binds the universe together in uniform law; Lagrange, Laplace, Leibnitz with their wondrous mathematical harmonies; Coulomb measuring out electricity ... Faraday, Ohm, Ampère, Joule, Maxwell, Hertz, Röntgen; and in another branch of science, Cavendish, Davy, Dalton, Dewar; and in another, Darwin, Mendel, Pasteur, Lister, Sir Ronald Ross. All these and many others, and some whose names have no memorial, form a great host of heroes, an army of soldiers – fit companions of those of whom the poets have sung ... There is the great Newton at the head of this list comparing himself to a child playing on the seashore gathering pebbles, whilst he could see with prophetic vision the immense ocean of truth yet unexplored before him ...

How often did you hear that sort of thing in a religious service? Or this, his gentle indictment of mindless patriotism, delivered on Empire Day at the close of the First World War? He went right

---

*'Public schools' are, as you might imagine, private schools! Only relatively affluent parents can afford them, which puts them at the opposite end of the political spectrum from the government-run Comprehensive schools (not invented in Sanderson's time) where education is free.

through the Sermon on the Mount, concluding each Beatitude with a mocking 'Rule Britannia'.

> Blessed are they that mourn, for they shall be comforted. Rule Britannia!
> Blessed are the meek, for they shall inherit the Earth. Rule Britannia!
> Blessed are the peacemakers, for they shall be called the children of God. Rule Britannia!
> Blessed are they that have been persecuted for righteousness sake. Rule Britannia!
> Dear souls! My dear souls! I wouldn't lead you astray for anything.

Sanderson's passionate desire to give the boys freedom to fulfil themselves would have thrown Health and Safety into a hissy fit, and set today's lawyers licking their chops with anticipation. He directed that the laboratories should be left unlocked at all times, so that boys could go in and work at their own research projects, even if unsupervised. The more dangerous chemicals were locked up, 'but enough was left about to disturb the equanimity of other masters who had less faith than the Head in that providence which looks after the young.' The same open door policy applied to the school workshops, the finest in the country, filled with advanced machine tools which were Sanderson's pride and joy. Under these conditions, one boy damaged a 'surface plate' by using it as an anvil against which to hammer a rivet. The culprit tells the story in *Sanderson of Oundle*:

> That did disconcert the Head for a little when it was discovered.* But my punishment was quite Oundelian. I had to make a study of the manufacture and use of surface plates and bring a report and explain it all to him. And after that I found I had learnt to look twice at a fine piece of work before I used it ill.

Incidents like this led eventually, and not surprisingly, to the workshops and laboratories again being locked when there was no adult supervision. But some boys felt the deprivation keenly and, in true Sandersonian fashion, they set out, in the workshops and the library (another of Sanderson's personal prides) to make an intensive study of locks.

*As well it might, for a 'surface plate' is a precisely machined plane surface, used for judging the flatness of objects.

In our enthusiasm we made skeleton keys for all Oundle, not only for the laboratories but for private rooms as well. For weeks we used the laboratories and workshops as we had grown accustomed to use them, but now with a keen care of the expensive apparatus and with precautions to leave nothing disorderly to betray our visits. It seemed that the Head saw nothing; he had a great gift for assuming blindness – until Speech Day came round, and then we were amazed to hear him, as he beamed upon the assembled parents, telling them the whole business, 'And what do you think my boys have been doing now?'

Sanderson's hatred of any locked door which might stand between a boy and some worthwhile enthusiasm symbolized his whole attitude to education. A certain boy was so keen on a project he was working on that he used to steal out of the dormitory at 2 a.m. to read in the (unlocked, of course) library. The Headmaster caught him there, and roared his terrible wrath for this breach of discipline (he had a famous temper and one of his maxims was 'Never punish except in anger'). Again, the boy himself tells the story:

The thunderstorm passed. 'And what are you reading, my boy, at this hour?' I told him of the work that had taken possession of me, work for which the day time was all too full. Yes, yes, he understood that. He looked over the notes I had been taking and they set his mind going. He sat down beside me to read them. They dealt with the development of metallurgical processes, and he began to talk to me of discovery and the values of discovery, the incessant reaching out of men towards knowledge and power, the significance of this desire to know and make and what we in the school were doing in that process. We talked, he talked for nearly an hour in that still nocturnal room. It was one of the greatest, most formative hours in my life ... 'Go back to bed, my boy. We must find some time for you in the day for this.'

I don't know about you, but that story brings me close to tears.

Far from coveting garlands in league tables by indulging the high-flyers,

Sanderson's most strenuous labours were on behalf of the average, and specially the 'dull' boys. He would never admit the word: if a boy was dull it was because he was being forced in the wrong direction, and he would make endless experiments to find how to get his interest ... he knew every boy by name and had a complete mental picture of his ability and character ... It was not enough that the majority should do well. 'I never like to fail with a boy.'

In spite of – perhaps because of – Sanderson's contempt for public examinations, Oundle did well in them. A faded, yellowing newspaper cutting dropped out of my second-hand copy of Wells's book:

In the higher certificates of the Oxford and Cambridge School examinations Oundle once again leads, having 76 successes. Shrewsbury and Marlborough tie for second place at 49 each.

Sanderson died in 1922, after struggling to finish a lecture to a gathering of scientists, at University College, London. The chairman, H. G. Wells himself, had just invited the first question from the floor when Sanderson dropped dead on the platform. The lecture had not been intended as a valediction, but the eye of sentiment can read the published text as Sanderson's educational testament, a summation of all he had learned in 30 years as a supremely successful and deeply loved headmaster.

My head ringing with the last words of this remarkable man, I closed the book and travelled on to University College, London, site of his swan song and my own modest speech to the conference of science teachers.

My subject, under the chairmanship of an enlightened clergyman, was evolution. I offered an analogy which teachers might use to bring home to their pupils the true antiquity of the universe. If a history were written at a rate of one century per page, how thick would the book of the universe be? In the view of a Young Earth Creationist, the whole history of the universe, on this scale, would fit comfortably into a slender paperback. And the scientific answer to the question? To accommodate all the volumes of history on the same scale, you'd need a bookshelf ten miles long. That gives the order of magnitude of the yawning gap between true science on the one hand, and the creationist teaching favoured by some schools on the other. This is not some disagreement of scientific detail. It is the difference between a

67

single paperback and a library of a million books. What would have offended Sanderson about teaching the Young Earth view is not just that it is false but that it is petty, small-minded, parochial, unimaginative, unpoetic and downright *boring* compared to the staggering, mind-expanding truth.

After lunching with the teachers I was invited to join their afternoon deliberations. Almost to a man and woman, they were deeply worried about the A-level syllabus and the destructive effects of exam pressure on true education. One after another, they came up to me and confided that, much as they would like to, they didn't *dare* to do justice to evolution in their classes. This was not because of intimidation by fundamentalist parents (which would have been the reason in parts of America). It was simply because of the A-level syllabus. Evolution gets only a tiny mention, and then only at the end of the A-level course. This is preposterous, for, as one of the teachers said to me, quoting the great Russian American biologist Theodosius Dobzhansky (a devout Christian, like Sanderson), 'Nothing in biology makes sense except in the light of evolution.'

Without evolution, biology is a collection of miscellaneous facts. Before they learn to think in an evolutionary way, the facts that the children learn will just be facts, with no binding thread to hold them together, nothing to make them memorable or coherent. With evolution, a great light breaks through into the deepest recesses, into every corner, of the science of life. You understand not only what is, but why. How can you possibly teach biology unless you *begin* with evolution? How, indeed, can you call yourself an educated person, if you know nothing of the Darwinian reason for your own existence? Yet, time and again, I heard the same story. Teachers had wanted to introduce their pupils to life's central theorem, only to be glottal-stopped dead in their tracks: 'Is that on my syllabus? Will it come up in my exam?' Sadly, they had to admit that the answer was no, and returned to the rote learning of disconnected facts as required for A-level success.

Sanderson would have hit the roof:

> I agree with Nietzsche that 'The secret of a joyful life is to live dangerously.' A joyful life is an active life – it is not a dull static state of so-called happiness. Full of the burning fire of enthusiasm, anarchic, revolutionary, energetic, daemonic, Dionysian, filled to overflowing with the terrific urge to create – such is the life of the man who risks safety and happiness for the sake of growth and happiness.

His spirit lived on at Oundle. His immediate successor, Kenneth Fisher, was chairing a staff meeting when there was a timid knock on the door and a small boy came in: 'Please, sir, there are Black Terns down by the river.' 'This can wait,' said Fisher decisively to the assembled committee. He rose from the Chair, seized his binoculars from the door and cycled off in the company of the small ornithologist, and – one can't help imagining – with the benign, ruddy-faced ghost of Sanderson beaming in their wake. Now *that's* education – and to hell with your league table statistics, your fact-stuffed syllabuses and your endless roster of exams.

That story of Fisher was told by my own inspiring Zoology teacher, Ioan Thomas, who had applied for the job at Oundle specifically because he admired the long-dead Sanderson and wanted to teach in his tradition. Some 35 years after Sanderson's death, I recall a lesson about *Hydra*, a small denizen of still fresh-water. Mr Thomas asked one of us, 'What animal eats Hydra?' The boy made a guess. Non-committally, Mr Thomas turned to the next boy, asking him the same question. He went right round the entire class, with increasing excitement asking each one of us by name, 'What animal eats Hydra? What animal eats Hydra?' And one by one we guessed. By the time he had reached the last boy, we were agog for the true answer. 'Sir, sir, what animal *does* eat Hydra?' Mr Thomas waited until there was a pin-dropping silence. Then he spoke, slowly and distinctly, pausing between each word.

> I don't know … (*Crescendo*) I don't know … (*Molto crescendo*) And I don't think Mr Coulson knows either. (*Fortissimo*) Mr Coulson! Mr Coulson!

He flung open the door to the next classroom and dramatically interrupted his senior colleague's lesson, bringing him into our room. 'Mr Coulson, do you know what animal eats Hydra?' Whether some wink passed between them I don't know, but Mr Coulson played his part well: he didn't know. Again the fatherly shade of Sanderson chuckled in the corner, and none of us will have forgotten that lesson. What matters is not the facts but how you discover and think about them: education in the true sense, very different from today's assessment-mad exam culture.

Sanderson's tradition that the whole school, not just the choir, even the tone deaf, should rehearse and bellow a part in the annual oratorio, also survived him, and has been widely imitated by other schools. His most famous innovation, the Week in Workshops (a full week for every pupil in every term, with all other work suspended) has not survived, but it was still going during my time in the fifties. It was eventually killed by exam pressure – of course – but a wonderfully Sandersonian phoenix has risen from its ashes. The boys, and now girls I am delighted to say, work out of school hours to build sports cars (and off-road go-carts) to special Oundle designs. Each car is built by one pupil, with help of course, especially in advanced welding techniques. When I visited Oundle last week, I met two overalled young people, a boy and a girl, who had recently left the school but had been welcomed back from their separate universities to finish their cars. More than 15 cars have been driven home by their proud creators during the past three years.

So Mr Sanderson, dear soul, you have a stirring, a light breeze of immortality, in the only sense of immortality to which the man of reason can aspire. Now let's whip up a gale of reform through the country, blow away the assessment-freaks with their never-ending cycle of demoralizing, childhood-destroying examinations, and get back to true education.

2

LIGHT WILL BE THROWN

The title of this section – and of its first chapter – is a quotation from the *Origin of Species*. Darwin was talking about light being thrown on human origins and he made it come true in his *Descent of Man*, but I like to think of all the other light that his ideas have thrown in so many different fields. Indeed, it was our second choice for the title for the whole book. The first essay in the section, **Light Will Be Thrown** (2.1), is the Foreword that I wrote very recently for a new student edition of *The Descent*, published by Gibson Square Books. In the course of writing it I discovered that Darwin was even more far-sighted than I had previously realized.

Darwin Triumphant (2.2) was my contribution to the second *Man and Beast* symposium, in Washington DC, 1991, with the subtitle 'Darwinism as a Universal Truth'. The phrase Universal Darwinism was one that I had introduced at the 1982 Cambridge conference to commemorate the centenary of Darwin's death. Darwinism is not just something that happens to be the basis of life on this planet. A good case can be made that it is fundamental to life itself, as a universal phenomenon wherever life may be found. If this is right, Darwin's light is thrown farther than was ever dreamed by that gentle and modest man.

One place where light could be thrown with advantage is the murky underworld of creationist propaganda. Television producers have such obvious power in the editing suite and the cutting room, it is amazing how seldom they abuse it. Tony Benn, the veteran socialist Member of Parliament, is said to switch on his own tape recorder, as a witness of potential foul play, whenever he is interviewed. Surprisingly, I have seldom found this necessary, and the only time I have ever been deliberately deceived was by an Australian creationist. How this disreputable story prompted me to publish

The 'Information Challenge' (2.3) is explained in the piece itself.

'A devil, a born devil, on whose nature, Nurture can never stick.' Gratified as Shakespeare might be to know how many of his lines have assumed household familiarity, I suspect that he might squirm at the modern over-exposure of the nature/nurture cliché. A flurry of publicity in 1993 for a so-called 'gay gene' on the X chromosome led to an invitation from the *Daily Telegraph* to expose the myths of 'genetic determinism'. The result was the piece reproduced here as **Genes Aren't Us** (2.4).

My literary agent John Brockman has the charisma to persuade his clients and others to drop everything and contribute to books of his own editing, even in the teeth of the better commercial judgement he might normally advise them to deploy. The distinction of his guest list flatters them in through the door of his salon (http://www.edge.org/) and before they know where they are they are correcting the proofs for a printed spin-off. **Son of Moore's Law** (2.5) was my futurological contribution to a typically fascinating on-line symposium, *The Next Fifty Years*.

Light Will Be Thrown

Foreword to a new Student Edition
of Darwin's *Descent of Man*[27]

Humanity is the missing guest at the feast of *The Origin of Species*. The famous 'Light will be thrown on the origin of man and his history' is a calculated understatement matched, in the annals of science, only by Watson and Crick's 'It has not escaped our notice that the specific pairing we have postulated immediately suggests a possible copying mechanism for the genetic material.' By the time Darwin finally got around to throwing that light in 1871, others had been there before him. And the greater part of *The Descent of Man* is not about humans but about Darwin's 'other' theory, sexual selection.

The Descent of Man was conceived as a single book but ended up as three, two of them bound together under the same title, with the second topic signalled by the subtitle, *Selection in Relation to Sex*. The third was *The Expression of the Emotions*, not my concern here, but Darwin tells us that it grew out of the original *Descent*, and he began writing it immediately after finishing *Descent*. Given that the idea of splitting the book was in Darwin's mind, it is at first sight surprising that he didn't spin off sexual selection as well. It would have seemed natural to publish chapters 8 to 18 as *Selection in Relation to Sex* followed by a second book, *The Descent of Man*, consisting of the present Chapters 1 to 8, and 19 to 21. That's a neat split into eleven chapters for each book, and many have wondered why he did not do this. I shall follow the same order – sexual selection followed by the descent of man – and then return at the end to the question of whether the two might have been split. In addition to discussing Darwin's book, I shall try to give some pointers to where the subject is moving today.

The ostensible connection between sexual selection and the descent of man is that Darwin believed the first was a key to understanding the second; especially to understanding human races, a topic which preoccupied Victorians more than it does us. But, as the historian and philosopher of science Michael Ruse has remarked to me, there was a tighter thread binding the two topics. They were the only two sources of disagreement between Darwin and his co-discoverer of natural selection. Alfred Russel Wallace never took kindly to sexual selection, at least in its full-blooded Darwinian form. And Wallace, though he coined the word Darwinism and described himself as 'more Darwinian than Darwin', stopped short of the materialism implied by Darwin's view of the human mind. These disagreements with Wallace were all the more important to Darwin because these two great men agreed on almost everything else. Darwin himself said, in a letter to Wallace of 1867:[28]

> The reason of my being so much interested just at present about sexual selection is, that I have almost resolved to publish a little essay on the origin of Mankind, and I still strongly think (though I failed to convince you, and this, to me is the heaviest blow possible) that sexual selection has been the main agent in forming the races of man.

The Descent of Man and *Selection in Relation to Sex* could be seen, then, as Darwin's two-pronged answer to Wallace. But it's also possible – and anyone who reads those chapters would forgive him – that he just got carried away by his enthusiasm for sexual selection.

The disagreements between Darwin and Wallace over sexual selection have been teased out by the Darwinian philosopher and historian Helena Cronin in her stylish book *The Ant and the Peacock*.[29] She even follows the two threads to the present day, classifying later theorists of sexual selection as 'Wallaceans' and 'Darwinians'. Darwin rejoiced in sexual selection. The naturalist in him loved the extravagant ostentation of stag beetles and pheasants, while the theorist and teacher knew that survival is only a means to the end of reproduction. Wallace could not stomach aesthetic whim as a sufficient explanation for the evolution of bright colours and the other conspicuous features for which Darwin invoked female (or in a few species male) choice. Even when persuaded that certain male features have evolved as advertisements aimed at females, Wallace insisted that the qualities they advertise must be utilitarian qualities. Females choose males not because they are pretty but because they are good providers, or something equally worthy. Modern Wallaceans such as William Hamilton[30] and Amotz Zahavi[31] see bright colours and other sexually selected advertisements as honest and uncheatable badges of true quality: health, for example, or resistance to parasites.

Darwin would have no problem with that, but he also was prepared to countenance pure aesthetic whim as a selective force in nature. Something about the female brain just likes bright coloured feathers, or whatever is the species equivalent, and that is a sufficient pressure for males to have evolved them, even if this is disadvantageous to the male's own survival. It was that leader among twentieth-century Darwinians, R. A. Fisher, who put the idea on a sound theoretical foundation by suggesting that female preference could be under genetic control and therefore subject to natural selection, in just the same way as the male qualities preferred.[32] The interaction between selection on female preference genes (inherited by both sexes) and simultaneously on male advertisement genes (also inherited by both sexes) provides the coevolutionary driving force for the expansion of ever more extravagant sexual advertisements. I suspect that Fisher's elegant reasoning, supplemented by more recent theorists such as R. Lande, might have reconciled Wallace to Darwin, because Fisher did not leave female whim unexplained, as an arbitrary given. The key point is that female whims of the future agree with those inherited from the past.[33]

The divide between Darwinian and Wallacean sexual selection, then, is one thing to bear in mind while reading the substantial middle section of *The Descent of Man*. Another is that Darwin made a clear distinction between sexual and natural selection, one which today is not always understood. Sexual selection is all about competition between members of the same sex for the opposite sex. It usually produces adaptations in males for out-competing other males: either for fighting males or for attracting females. It does not include all the rest of the apparatus of sexual reproduction. A penis, in its capacity as an organ of intromission, is a manifestation of natural selection, not sexual selection. A male needs a penis to reproduce, whether or not competing males are around. But male vervet monkeys (*Cercopithecus aethiops*) have a bright red penis set off by a sky-blue scrotum, which together are shown off in dominance displays to other males. It is for their colours, not the organs themselves, that Darwin would invoke sexual selection.

To decide whether something is a sexually selected adaptation or not, do the following thought experiment. Imagine that all competitors of the same sex could somehow be magicked away. If the pressure for the adaptation now disappears, it was sexually selected. In the case of the vervet monkeys it is reasonable to guess, as Darwin surely would, that if competition from rival males were removed by a magic wand, the penis and scrotum would remain, but their red and blue colour scheme would fade. The ornate colours are a product of sexual selection, the utilitarian organs of sperm production and intromission are manifestations of natural selection. Darwin would have loved the baroque and spiky penises documented by W. G. Eberhard in his book, *Sexual Selection and Animal Genitalia*.[34]

The distinguished American philosopher Daniel Dennett has credited Darwin with the greatest idea ever to occur to a human mind.[35] This was natural selection, of course, and I would include sexual selection as part of the same idea. But Darwin was not only a deep thinker, he was a naturalist of encyclopaedic knowledge and (which by no means necessarily follows) had the ability to hold it in his head and deploy it in constructive directions. He was a master encyclopaedist, who collated huge quantities of information and observations solicited from naturalists all around the world, each gentleman meticulously acknowledged for having 'attended to' the subject and sometimes complimented as a 'reliable observer'. I find an addictive fascination in his Victorian prose style, quite apart from the feeling one gets of having been ushered into the presence of one of the great minds of all time.

Prescient as he was (Michael Ghiselin has said that he worked at least a century ahead of his time[36]) Darwin was still a Victorian, and his book must be read in the context of its age, warts and all. What will grate most irksomely on the modern ear is the unquestioned Victorian presumption that animals in general, and humans in particular, are disposed on a ladder of increasing superiority. Like all Victorians, Darwin happily referred to particular species as 'lowly in the scale of nature'. Even some modern biologists do this, though they should not, for all living species are cousins who have been evolving for exactly the same length of time since the common ancestor.[37] What educated moderns never do, but equivalent Victorians always did, is think of human races in the same hierarchical way. It requires a special effort for us to read something like the following without distaste:

It seems at first sight a monstrous supposition that the jet blackness of the negro has been gained through sexual selection [i.e. is attractive to the opposite sex] ... The resemblance of *Pithecia satanas* with his jet black skin, white rolling eyeballs, and hair parted on the top of the head, to a negro in miniature, is almost ludicrous.[38]

It is a mark of historical infantilism to view the writings of one century through the politically tinted glasses of another. The very title, *Descent of Man*, will raise hackles among those naively locked into the mores of our own time. It can be argued that reading historic documents that violate the taboos of one's own century gives valuable lessons in the ephemerality of such mores. Who knows how our descendants will judge us?

Less obvious, but as important to understand, are the changes in the scientific climate. In particular, it is hard to overstate the fact that Darwin's genetics were pre-Mendelian. The intuitively plausible blending inheritance theory of his time was not just wrong, it was grievously wrong and especially grievous for natural selection. Darwinism's incompatibility with blending inheritance was pointed out in a hostile review of the *Origin* by the Scottish engineer Fleeming Jenkin. Variation tends to disappear with every blending generation, leaving not enough for natural selection to get its teeth into. What Jenkin should have realized is that blending inheritance is incompatible not just with Darwinian theory but with obvious fact. If it were really true that variation disappeared, every generation should be more uniform than the previous one. By now, all individuals should be as indistinguishable as clones. Darwin needed only to retort to Jenkin: Whatever the reason, it is obviously the case that there is plenty of inherited variation and that's good enough for my purposes.

It is often claimed that the answer to the riddle lay on Darwin's shelves, in the uncut pages of the proceedings of the Brunn Natural History Society, where nestled Gregor Mendel's paper on *Versuche über Pflanzen-Hybriden*. Unfortunately this poignant story seems to be an urban myth. The two scholars best placed (at Cambridge and at Down House) to know what was in Darwin's personal library can find no evidence that he ever subscribed to the proceedings, nor does it seem likely that he would have done so.[39] They have no idea where the legend of the 'uncut pages' originated. Once originated, however, it is easy to see that its very poignancy might speed its proliferation. The whole affair would make a nice little project in memetic research, complementing that other popular urban legend, the agreeable falsehood that Darwin turned down an offer from Marx to dedicate *Das Kapital* to him.[40]

Mendel did indeed have exactly the insight Darwin needed. The relationship of his work to the Jenkin critique, however, would not have been immediately obvious to the Victorian mind. Even after Mendel's work was rediscovered in 1900 and inspired the Hardy-Weinberg Law in 1908, it was not until Fisher came along in 1930* that its supreme relevance to Darwinism was widely understood. If heredity is particulate, variation does not disappear but is reconstituted in every generation. Neo-Darwinian evolution precisely means change in gene frequencies in gene pools. What is genuinely poignant is that Darwin himself came tantalizingly close. Fisher[41] quotes him in a letter to Huxley of 1857:

> I have lately been inclined to speculate, very crudely and indistinctly, that propagation by true fertilization will turn out to be a sort of mixture, and not true fusion, of two distinct individuals, or rather of innumerable individuals, as each parent has its parents and ancestors. I can understand on no other view the way in which crossed forms go back to so large an extent to ancestral forms. But all this, of course, is infinitely crude.

Fisher cleverly remarked that Mendelism has a kind of necessary

---

*Actually rather earlier, but 1930 was when Fisher published his landmark book.

plausibility which could have led to its discovery by any thinker in a mid-Victorian armchair (quoted on page 95). He might have added that particulate inheritance stares us in the face whenever we contemplate sex itself (as we not infrequently do). All of us have one female and one male parent, yet each of us is either male or female, not an intermediate hermaphrodite. Fascinatingly, Darwin himself made this very point, clearly, in an 1866 letter to Wallace,[42] which Fisher would surely have quoted had he known of it.

> My dear Wallace … I do not think you understand what I mean by the non-blending of certain varieties. It does not refer to fertility; an instance will explain. I crossed the Painted Lady and Purple sweet peas, which are very differently coloured varieties, and got, even out of the same pod, both varieties perfect but none intermediate. Something of this kind I should think must occur at least with your butterflies & the three forms of Lythrum; tho' these cases are in appearance so wonderful, I do not know that they are really more so than every female in the world producing distinct male and female offspring …
>
> Believe me, yours very sincerely
>
> Ch. Darwin

Here Darwin comes closer to anticipating Mendel than in the passage quoted by Fisher, and he even mentions his own Mendel-like experiments on sweet peas. I am extremely grateful to Dr Seymour J. Garte of New York University, who found this letter by chance in a volume of correspondence between Darwin and Wallace in the British Library in London, immediately recognized its significance and sent a copy to me.

Another piece of Darwin's unfinished business later sorted out by Fisher was the matter of the sex ratio, and how it evolves under natural selection. Fisher begins by quoting the Second Edition of *The Descent of Man*, in which Darwin prudently said:

> I formerly thought that when a tendency to produce the two sexes in equal numbers was advantageous to the species, it would follow from natural selection, but I now see that the whole problem is so intricate that it is safer to leave its solution to the future.

Fisher's own solution[43] made no appeal to species advantage. Instead he pointed out that, since every individual born has one father and one mother, the total male contribution to posterity must equal the total female contribution. If the sex ratio is anything other than 50/50, therefore, an individual of the minority sex can expect, other things being equal, a greater share of descendants, and this will set up selection in favour of rebalancing the sex ratio. Fisher rightly used economic language to express the strategic decisions involved: they are decisions over how to allocate parental expenditure. Natural selection will favour parents who spend proportionately more food or other resources on offspring of the minority sex. Such correcting selection will continue until the total expenditure on sons in the population balances the total expenditure on daughters. This will amount to equal numbers of males and females, except in those cases where offspring of one sex cost more to rear than offspring of the other. If, for example, it costs twice as much food to rear a son than a daughter (perhaps to make sons big enough to compete effectively with rival males) the stable sex ratio will be twice as many females as males. This is because the strategic alternative to one son is not one daughter but two. Fisher's powerful logic has been extended and refined in various ways, for example by W. D. Hamilton[44] and E. L. Charnov[45].

Once again, and notwithstanding the quotation above from the Second Edition of *The Descent of Man*, Darwin himself, in the First Edition, came remarkably close to anticipating Fisher, although without the economic language of parental expenditure:

> Let us now take the case of a species producing, from the unknown causes just alluded to, an excess of one sex – we will say of males – these being superfluous and useless, or nearly useless. Could the sexes be equalized through natural selection? We may feel sure, from all characters being variable, that certain pairs would produce a somewhat less excess of males over females than other pairs. The former, supposing the actual number of the offspring to remain constant, would necessarily produce more females, and would therefore be more productive. On the doctrine of chances a greater number of the offspring of the more productive pairs would survive; and these would inherit a tendency to procreate fewer males and more females. Thus a tendency toward equalization of the sexes would be brought about.

Sadly, Darwin deleted this remarkable passage when he came to prepare the Second Edition, preferring the more cautious paragraph later to be quoted by Fisher. Darwin's partial anticipation of Fisher in the First Edition of *Descent* is all the more impressive because, as Alan Grafen points out to me, Fisher's argument depends crucially on a fact which was not available to Darwin, namely that the two parents make an equal genetic contribution to every offspring. Indeed, in historical times, different schools of thought (the spermists and the ovists respectively) had held that the male, or the female sex had a monopoly on heredity.

The whole question of Fisher's sources for the sex ratio theory has been meticulously sleuthed by Professor A. W. F. Edwards of Cambridge University,[46] himself one of Fisher's most distinguished pupils. Edwards not only notes Darwin's priority over the essential argument and the odd fact that he deleted it from the Second Edition. He also shows how Darwin's argument was taken up and developed by a series of other workers whose writings were probably known to Fisher. First Carl Düsing of Jena, in 1884, reiterated and clarified Darwin's argument. Next, in 1908 the Italian statistician Corrado Gini discussed the argument more critically. Finally in 1914, the eugenicist J. A. Cobb gave a form of the argument which seems to have all the refinements of Fisher's own of 1930, including the economic idea of parental expenditure. Cobb seems to have been unaware of Darwin's priority, but Edwards is persuasive that Fisher was aware of Cobb's. Edwards remarks that:

> commentators have assumed, and most have firmly stated, that the argument was original to Fisher, though he did not claim it to be, nor did he refer to it either before or after 1930 in any of his other publications. Indeed, there is no evidence that he saw it as particularly novel, remarkable, or likely to lead to major developments in evolutionary biology … he may well have regarded it as public property by 1930.

Edwards himself is one of those (I am another) who once overlooked the crucial difference between the First and Second Editions of *The Descent*.

Fisher's economic view of sex was developed further by Robert L. Trivers, writing in a volume published to commemorate the centenary of *The Descent of Man*.[47] Trivers's subtle application of the theory of parental investment (his name for what Fisher had called parental expenditure) to male and female roles in sexual selection greatly illuminates the facts collected by Darwin in the middle chapters of *Descent*. Trivers defines parental investment (PI) as (what economists would call) an opportunity cost. The cost to a parent of investing in a particular child is measured in correspondingly lost opportunity to invest in others, present or future. Sexual inequality is fundamentally economic. The mother typically invests more in any individual offspring than the father does, and this inequality has far-reaching consequences, which reach even further in a kind of self-feeding process. A member of the low-investing sex (usually male) who persuades a member of the high-investing sex (usually female) to mate with him has gained an economic prize worth fighting (or otherwise competing) for. This is why males typically devote more effort to competing with other males, while females typically shunt their effort away from competing with other females and into investing in offspring. It is why, when one sex is more brightly coloured than the other, it is typically the male. It is why, when one sex is more choosy in selecting a mate, it is typically the female. And it is why variance in reproductive success is typically higher among males than among females: the most successful male may have many times more descendants than the least successful male, where the most successful female is only somewhat more successful than the least successful female. The Fisher/ Trivers economic inequalities between the sexes should be kept in mind while reading Darwin's enthralling review of sexual selection through the animal kingdom. It is a most striking example of a single idea uniting and explaining, at one blow, a multitude of seemingly disparate facts.

Now, to the descent of man itself. Darwin's guess that our species arose in Africa was typically ahead of its time, amply confirmed today by numerous fossils, none of which was available to him. We are African apes, closer cousins to chimpanzees and gorillas than they are to orang utans and gibbons, let alone monkeys. Darwin's 'quadrumana' were defined so as to exclude humans: they were all the apes and monkeys, with a hand bearing an opposable digit on the hindlegs as well as the forelegs. The early chapters of his book are concerned to narrow the perceived gap between ourselves and the quadrumana, a gap which Darwin's target audience would have seen as yawning between the top rung of a ladder and the next rung down. Today we would not (or should not) see a ladder at all. Instead, we should hold in our minds the branching tree diagram which is the only illustration in *The Origin of Species*. Humanity is just one little twig, nestling among many others somewhere in the middle of a thicket of African apes.

Two vital techniques which were unavailable to Darwin are radioactive dating of rocks, and molecular evidence including the 'molecular clock'. Where Darwin, in his quest to demonstrate the similarity between ourselves and the quadrumana, could point to comparative anatomy supplemented by charming anecdotes of psychological and emotional resemblance (arguments extended in *The Expression of the Emotions*), we are privileged to know the exact letter-by-letter sequence of massive DNA texts. It is claimed that more than 98 per cent of the human genome, when measured in this way, is identical with chimpanzees'. Darwin would have been spellbound. Such closeness of resemblance, and such precision in measuring it, would have delighted him beyond his dreams.

Nevertheless, we must beware of being carried away by the euphoria of it all. That 98 per cent doesn't mean we are 98 per cent chimpanzees. And it really matters which unit you choose to make your comparison. If you count the number of whole genes that are identical, the figure for humans and chimpanzees would be close to zero. This is not a paradox. Think of the human genome and the chimpanzee genome as two editions of the same book, say the first and second editions of *The Descent of Man*. If you count the number of letters that are identical to their opposite numbers in the other edition, it is probably well over 90 per cent. But if you count the number of chapters that are identical, it may well be zero. This is because it takes only one letter to be different, anywhere in a chapter, for the whole chapter to be judged different between the two editions. When you are measuring the percentage similarity between two texts, whether two editions of a book or two editions of an African ape, the unit of comparison you choose (letter or chapter, DNA base pair or gene) makes a huge difference to the final percentage similarity.

The point is that we should use such percentages not for their absolute value but in comparisons between animals. The 98 per cent figure for humans and chimpanzees starts to make sense when we compare it with the 96 per cent resemblance between humans and orang utans (it is the same 96 per cent between chimpanzees and orang utans, and the same between gorillas and orang utans, because all the African apes are connected to the Asian orang utans via a shared African ancestor). For the same kind of reason, all the great apes share 95 per cent of their genomes with the gibbons and siamangs. And all the apes share 92 per cent of their genomes with all Old World monkeys.

The hypothesis of a molecular clock allows us to use such percentage figures to put a date on each of the splits in our family tree. It assumes that evolutionary change, at the molecular genetic level, proceeds at an approximately fixed rate for each gene. This is in accordance with the widely accepted neutral theory of the Japanese geneticist Motoo Kimura. Kimura's neutral theory is sometimes seen as anti-Darwinian but it is not. It is *neutral* with respect to Darwinian selection. A neutral mutation is one that makes no difference to the functioning of the protein produced. The post-mutation version is no better and no worse than the pre-mutation version, where both may be vital to the life of the organism.

From a Darwinian point of view, neutral mutations are not mutations at all. But from a molecular point of view they are extremely useful mutations because their fixed rate makes the clock reliable. The only point of controversy introduced by Kimura is how *many* mutations are neutral. Kimura thought it was the great majority which, if true, is very nice for the molecular clock. Darwinian selection remains the only explanation for adaptive evolution and it is arguable (I would argue) that most if not all of the evolutionary changes we actually see in the macroscopic world (as opposed to those concealed among the molecules) are adaptive and Darwinian.

As so far described, the molecular clock gives relative timings but not absolute ones. We can read off timings of evolutionary splits, but only in arbitrary units. Fortunately, in another great advance that would have entranced Darwin, various absolute clocks are available for dating fossils. These include the known rates of radioactive decay of isotopes in volcanic rocks sandwiching the sedimentary strata in which fossils are found. By taking a group of animals with a rich fossil record and dating the splits in their family tree two ways – by the molecular genetic clock and by radioactive clocks – the arbitrary units of the genetic clock can be validated, and simultaneously calibrated in real millions of years. This is how we can estimate that the split between humans and chimpanzees occurred between 5 and 8 million years ago, the split between African apes and orang utans about 14 million years ago, and the split between apes and Old World monkeys about 25 million years ago.

Fossils, all discovered after *Descent* was published, provide us with a sporadic picture of some possible intermediates connecting us to our common ancestor with chimpanzees. Unfortunately, there are no fossils connecting modern chimpanzees to that shared ancestor, but on our side of the split reports of new fossil finds are coming in at a rate which I find exciting and surely Darwin would have too. Going back in steps of roughly one million years we find: *Homo erectus*, *Homo habilis*, *Australopithecus afarensis*, *Australopithecus anamensis*, *Ardipithecus*, *Orrorin* and, a recent discovery which may date from as long ago as 7 million years, *Sahelanthropus*. That last find is from Chad, far to the west of the great Rift Valley which had hitherto been thought to constitute a geographic barrier dividing our lineage from that of the chimpanzees. It is good for our orthodoxies to be upset from time to time.

We must beware of assuming that this temporal series of fossils represents an ancestor/descendant series. It is always safer to assume that fossils are cousins rather than ancestors, but we need not be shy of guessing that earlier cousins may tell us at least something about the true ancestors among their contemporaries.

What are the main changes that occurred since our split from the chimpanzees? Some, such as our loss of body hair, are interesting, but fossils can tell us nothing directly about them. The two main changes that fossils can help us with, and where we therefore have a big advantage over Darwin, are that we rose up on our hind legs, and that our brains got rather dramatically bigger. Which of these changes came first, or did they happen together? All three views have been supported, and controversy has gone back and forth over the decades. Darwin thought the two big changes happened in concert, and he makes out a plausible case. But this is a rare instance where Darwin's tentative guess has turned out wrong. The fossils give a satisfyingly decisive and clear answer.[48] Bipedality came first, and its evolution was more or less complete before the brain started to swell. Three million years ago, *Australopithecus* was bipedal and had feet like ours, although it probably still retreated up trees. But its brain, relative to its body size, was the same size as a chimpanzee's, and presumably the same as the shared ancestor with chimpanzees. Nobody knows whether the bipedal gait set up new selection pressures that encouraged the brain to grow, but Darwin's original arguments for simultaneous evolution can be adapted to make that plausible. Perhaps the enlargement of the brain had something to do with language, but here nobody knows and disagreements abound. There is evidence that particular parts of the human brain are uniquely pre-wired to handle specific universals of language, although the particular language spoken is, of course, locally learned.[49]

Another twentieth-century idea which is probably important in human evolution, and which again would have intrigued Darwin, is neoteny: evolutionary infantilization. The axolotl, an amphibian living in a Mexican lake, looks just like the larva of a salamander, but it can reproduce, and has chopped off the adult, salamander stage of the life history. It is a sexually mature tadpole. Such neoteny has been suggested as a way in which a lineage can suddenly initiate an entirely new direction of evolution, at a stroke. Apes don't have a discrete larval stage like a tadpole or a caterpillar, but a more gradualistic version of neoteny can be discerned in human evolution. Juvenile chimpanzees resemble humans far more than adult chimpanzees do. Human evolution can be seen as infantilism. We are apes that became sexually mature while still morphologically juvenile.[50] If humans could live for 200 years, would we finally 'grow up', drop on all fours and develop huge prognathous chimpanzee-like jaws? The possibility has not been lost on writers of ironic fiction, notably Aldous Huxley in *After Many a Summer*. He presumably learned about neoteny from his elder brother Julian, who was one of the pioneers of the idea and did amazing research on axolotls, injecting hormones to make them turn into salamanders never before seen.

Let me end by bringing together once again the two halves of Darwin's book. He went to town on sexual selection in *The Descent of Man* because he thought it was important in human evolution, and especially because he thought it was the key to understanding the differences among human races. Race, in Victorian times, was not the political and emotional minefield it is today, when one can give offence by so much as mentioning the word. I shall tread carefully, but I cannot ignore the topic because it is prominent in Darwin's book and especially germane to the unification of its two parts.

Darwin, like all Victorians, was intensely aware of the differences among humans but he also, more than most of his contemporaries, emphasized the fundamental unity of our species. In *Descent* he carefully considered, and decisively rejected, the idea, rather favoured in his own time, that different human races should be regarded as separate species. Today we know that, at the genetic level, our species is more than usually uniform. It has been said that there is more genetic variation among the chimpanzees of a small region of Africa than among the entire world population of humans (suggesting that we have been through a bottleneck in the past hundred thousand years or so). Moreover, the great majority of human genetic variation is to be found within races, not between them. This means that if you were to wipe out all human races except one, the great majority of human genetic variance would be preserved. The variance between races is just a bit extra, stuck on the top of the greater quantity of variation within all races. It is for this reason that many geneticists advocate the complete abandonment of the concept of race.

At the same time – the paradox is similar to one recognized by Darwin – the superficially conspicuous features characteristic of local populations around the world seem very different. A Martian taxonomist who didn't know that all human races happily interbreed with one another, and didn't know that most of the underlying genetic variance in our species is shared by all races, might be tempted by our regional differences in skin colour, facial features, hair, body size and proportions to split us into more than one species. What is the resolution of the paradox? And why did such pronounced superficial differences evolve in different geographical areas, while most of the less conspicuous variation is dotted around across all geographical areas? Could Darwin have been right all along? Is sexual selection the answer to the paradox? The distinguished biologist Jared Diamond thinks so,[51] and I am inclined to agree.

Utilitarian answers have been suggested to the question of the evolution of racial differences, and there may well be some truth in them. Dark skin may protect against skin cancer in the tropics, light skin admit beneficial rays in sun-starved latitudes where there is a danger of Vitamin D deficiency. Small stature probably is of benefit to hunters in dense forest, such as the pygmies of central Africa, and various independently evolved hunter gatherers of Amazon and South East Asian forests. The ability to digest milk when adult seems to have evolved in peoples who, for cultural reasons, prolong the use of this primitively juvenile food. But I am impressed by the diversity of features that are superficial and conspicuous, while deeper differences are so slight.

What sexual selection explains, better than natural selection, is diversity that seems arbitrary, even driven by aesthetic whim. Especially if the variation concerned is geographical. And also especially if some of the features concerned, for example beards and the distribution of body hair and subcutaneous fat deposits, differ between the sexes. Most people have no problem in accepting an analogue of sexual selection for culturally mediated fashions like headdresses, body paint, penis sheaths, ritual mutilations or ornamental clothes. Given that cultural differences such as those of language, religion, manners and customs certainly provide resistance to interbreeding and gene flow, I think it is entirely plausible that genetic differences between peoples of different regions, at least where superficial, externally prominent features are concerned, have evolved through sexual selection. Our species really does seem to have unusually conspicuous, even ostentatious, superficial differences between local populations, coupled with unusually low levels of overall genetic variation. This double circumstance carries, to my mind, the stamp of sexual selection.

In this respect, human races seem a lot like breeds of dog,[52] another favourite topic of Darwin. Superficially, the domestic breeds of dogs are astonishingly varied, even more so than human races, yet the underlying genetic differences are slight, and they are all clearly descended from wolves within the past few thousand years.[53] Reproduction isolation is today maintained by disciplined pedigree breeders, and the shapes and colours of the dogs themselves are steered through their rapid evolution by the whim of the human eye rather than the whim of female dogs. But the essential features of the situation, as Darwin realized, are similar to those of sexual selection.

In this, as in so much else, I suspect that Darwin was right. Sexual selection really is a good candidate for explaining a great deal about the unique evolution of our species. It may also be responsible for some unique features of our species which are shared equally by all races, for example our enormous brain. Geoffrey Miller, in *The Mating Mind*,[54] has strongly developed precisely this case, and Darwin would have loved it no less because Miller takes a Wallacean view of sexual selection. It is starting to look as though, despite initial appearances, Darwin really was right to bring together, in one volume, *Selection in Relation to Sex* and *The Descent of Man*.

2.2

Darwin Triumphant[55]
Darwinism as a Universal Truth

If we are visited by superior creatures from another star system – they will have to be superior if they are to get here at all – what common ground shall we find for discussion with them? Shall we overcome the barriers simply by learning one another's language, or will the subjects that interest our two cultures be so divergent as to preclude serious conversation? It seems unlikely that the star travellers will want to talk about many of our intellectual stocks-in-trade: about literary criticism or music, religion or politics. Shakespeare may mean nothing to those without human experiences and human emotions, and if they have a literature or an art these will probably be too alien to excite our sensibilities. To name two thinkers who have more than once been promoted as Darwin's equals, I rather doubt whether our visitors will have much interest in talking about Marx or Freud, other than perhaps as anthropological curiosities. We have no reason to suppose that these men's works are of more than local, parochial, human, earthly, post-Pleistocene (some would add European and male) significance.

Mathematics and physics are another matter. Our guests may find our level of sophistication quaintly low, but there will be common ground. We shall agree that certain questions about the universe are important, and we shall almost certainly agree on the answers to many of these questions. Conversation will flourish, even if most of the questions flow one way and most of the answers the other. If we discuss the histories of our respective cultures, our visitors will surely point with pride, however far back in time, to their equivalents of Einstein and Newton, of Planck and Heisenberg. But they won't point to an equivalent of Freud or Marx any more than we, visiting a hitherto undiscovered tribe in a remote forest clearing, would nominate our civilization's equivalent of the local rainmaker or gully-gully man. One does not have to disparage the local achievements of Freud and Marx on this planet to agree that their findings have no universality.

What about Darwin? Will our guests revere another Darwin as one of their greatest thinkers of all time? Shall we be able to have a serious conversation with them about evolution? I suggest that the answer is yes (unless, as a colleague suggests to me, their Darwin is on the expedition and we are her Galapagos*). Darwin's achievement, like Einstein's, is universal and timeless, whereas that of Marx is parochial and ephemeral. That Darwin's *question* is universal, wherever there is life, is surely undeniable. The feature of living matter that most demands explanation is that it is almost unimaginably complicated in directions that convey a powerful illusion of deliberate design. Darwin's question, or rather the most fundamental and important of Darwin's many questions, is the question of how such complicated 'design' could come into being. All living creatures, everywhere in the universe and at any time in history, provoke this question. It is less obvious that Darwin's *answer* to the riddle – cumulative evolution by non-random survival of random hereditary changes – is universal. It is at first sight conceivable that Darwin's answer might be valid only parochially, only for the kind of life that happens to exist in our own little clearing in the universal forest. I have previously made the case that this is not so,[56] that the general form of Darwin's answer is not merely incidentally true of our kind of life but almost certainly true of all life, everywhere in the universe. Here, let me for the moment make the more modest claim that, at the very least, Darwin's bid for immortality is closer to the Einstein end of the spectrum than to the Marx end. Darwinism really matters in the universe.

When I was an undergraduate in the early nineteen sixties, we were taught that although Darwin was an important figure in his own time, modern neo-Darwinism was so much further advanced that it hardly deserved the name Darwinism at all. My father's

*This is how my friend worded her suggestion. The joke was rather ruined by the political scruples of the original article's copy-editor, who changed 'her Galapagos' to 'his or her Galapagos'.

generation of biologist undergraduates read, in an authoritative *Short History of Biology*[57], that

> ... the struggle of living forms leading to natural selection by the survival of the fittest, is certainly far less emphasized by naturalists now than in the years that immediately followed the appearance of Darwin's book. At the time, however, it was an extremely stimulating suggestion.

And the generation of biologists before that could read, in the words of William Bateson, perhaps the dominant British geneticist of the time,

> We go to Darwin for his incomparable collection of facts [but] ... for us he speaks no more with philosophical authority. We read his scheme of Evolution as we would those of Lucretius or Lamarck ... The transformation of masses of populations by imperceptible steps guided by selection is, as most of us now see, so inapplicable to the fact that we can only marvel ... at the want of penetration displayed by the advocates of such a proposition.[58]

And yet the editors of this volume can commission an article with the title 'Darwin Triumphant'. I do not normally like writing to titles that others have proposed, but I can accept this one without reservation. In the last quarter of the twentieth century, it seems to me that Darwin's standing among serious biologists (as opposed to nonbiologists influenced by religious preconceptions) is rightly as high as it has been at any time since his death. A similar story, of even more extreme eclipse in earlier years followed by triumphant recent rehabilitation, can be told of Darwin's 'other theory', that of sexual selection.*

It is only to be expected that, a century and a quarter on, the version of his theory that we now have should be different from the original. Modern Darwinism is Darwinism plus Weismannism plus Fisherism plus Hamiltonism (arguably plus Kimuraism and a few other *isms*). But when I read Darwin himself, I am continually astonished at how modern he sounds. Considering how utterly wrong he was on the all-important topic of genetics, he showed

*See 'Light Will Be Thrown' (pp. 73-90).

an uncanny gift for getting almost everything else right. Maybe we are neo-Darwinists today, but let us spell the *neo* with a very small *n*! Our neo-Darwinism is very much in the spirit of Darwin himself. The changes that Darwin would see if he came back today are in most cases changes that, I venture to suggest, he would instantly approve and welcome as the elegant and obviously correct answers to riddles that troubled him in his own time. Upon learning that evolution is change in *frequencies* within a pool of *particulate* hereditary elements, he might even quote T. H. Huxley's alleged remark upon reading the *Origin* itself: 'How extremely stupid not to have thought of that!'*

I referred to Darwin's gift for getting things right, but surely this can only mean right as we see it today. Shouldn't we be humble enough to admit that our right may be utterly wrong in the sight of future scientific generations? No, there are occasions when a generation's humility can be misplaced, not to say pedantic. We can now assert with confidence that the theory that the Earth moves round the Sun not only is right in our time but will be right in all future times even if flat-Earthism happens to become revived and universally accepted in some new dark age of human history. We cannot quite say that Darwinism is in the same unassailable class. Respectable opposition to it can still be mounted, and it can be seriously argued that the current high standing of Darwinism in educated minds may not last through all future generations. Darwin may be triumphant at the end of the twentieth century, but we must acknowledge the possibility that new facts may come to light which will force our successors of the twenty-first century to abandon Darwinism or modify it beyond recognition. But is there, perhaps, an essential core of Darwinism, a core that Darwin himself might have nominated as the

*Of the two stories about Huxley that have become chestnuts, I greatly prefer this to the one about his so-called 'debate' with the Bishop of Oxford, Sam Wilberforce. There is something admirably honest about Huxley's exasperation at not having thought of such a simple idea. I have long found it a complete mystery why it had to wait until the nineteenth century before anyone thought of it. Archimedes' and Newton's achievements seem, on the face of it, far more difficult. But the fact that nobody *did* think of natural selection before the nineteenth century clearly shows that I am wrong. As does the fact that so many people, even today, don't get it.

irreducible heart of his theory, which we might set up as a candidate for discussion as potentially beyond the reach of factual refutation?

Core Darwinism, I shall suggest, is the minimal theory that evolution is guided in adaptively nonrandom directions by the nonrandom survival of small random hereditary changes. Note especially the words *small* and *adaptively*. *Small* implies that adaptive evolution is gradualistic, and we shall see why this must be so in a moment. *Adaptive* does not imply that all evolution is adaptive, only that core Darwinism's concern is limited to the part of evolution that is. There is no reason to assume that all evolutionary change is adaptive.[59] But even if most evolutionary change is not adaptive, what is undeniable is that enough of evolutionary change is adaptive to demand some kind of special explanation. It is the part of evolutionary change that *is* adaptive that Darwin so neatly explained. There could be any number of theories to explain nonadaptive evolution. Nonadaptive evolution may or may not be a real phenomenon on any particular planet (it probably is on ours, in the form of the large-scale incorporation of neutral mutations), but it is not a phenomenon that awakes in us an avid hunger for an explanation. Adaptations, especially complex adaptations, awake such a powerful hunger that they have traditionally provided one of the main motivations for belief in a supernatural Creator. The problem of adaptation, therefore, really was a big problem, a problem worthy of the big solution that Darwin provided.

R. A. Fisher[60] developed a case, which did not make any appeal to particular facts, for the armchair deducibility of Mendelism.

> It is a remarkable fact that had any thinker in the middle of the nineteenth century undertaken, as a piece of abstract and theoretical analysis, the task of constructing a particulate theory of inheritance, he would have been led, on the basis of a few very simple assumptions, to produce a system identical with the modern scheme of Mendelian or factorial inheritance.

Is there a similar statement that could be made about the inevitability of the core of Darwin's scheme of evolution by natural selection? Although Darwin and Wallace themselves were field naturalists who made extensive use of factual information to support their theory, can we now, with hindsight, argue that there should have been no need for the *Beagle*, no need for the Galapagos and Malay Archipelagos? Should any thinker, faced with the problem formulated in the right way, have been able to arrive at the solution – core Darwinism – without stirring from an armchair?

Part of core Darwinism arises almost automatically from the problem that it solves, if we express that problem in a particular way, as one of mathematical search. The problem is that of finding, in a gigantic mathematical space of all possible organisms, that tiny minority of organisms that is adapted to survive and reproduce in available environments. Again, Fisher put it with characteristically powerful clarity.

> An organism is regarded as adapted to a particular situation, or to the totality of situations which constitute its environment, only in so far as we can imagine an assemblage of slightly different situations, or environments, to which the animal would on the whole be less well adapted; and equally only in so far as we can imagine an assemblage of slightly different organic forms, which would be less well adapted to that environment.

Imagine some nightmarish mathematical menagerie in which is found the all but infinitely large set of conceivable animal forms that could be cobbled together by randomly varying all the genes in all genomes in all possible combinations. For brevity, although it is not as precise a phrase as its mathematical tone leads one to think, I shall refer to this as the set of all possible animals (fortunately the argument I am developing is an order-of-magnitude argument which does not depend on numerical precision). Most of the members of this ill-favoured bestiary will never develop beyond the single-cell stage. Of the very few that manage to be born (or hatch, etc.), most will be hideously misshapen monstrosities who will die early. The animals that actually exist, or have ever existed, will be a tiny subset of the set of all possible animals. Incidentally, I use *animal* purely for convenience. By all means substitute *plant* or *organism*.

It is convenient to imagine the set of all possible animals as arrayed in a multidimensional genetic landscape.* *Distance* in this landscape means genetic distance, the number of genetic changes that would have to be made in order to transform one animal into another. It is not obvious how one would actually compute the genetic distance between any two animals (because not all animals have the same number of genetic loci); but again the argument does not rely upon precision, and it is intuitively obvious what it means, for instance, to say that the genetic distance between a rat and a hedgehog is larger than the genetic distance between a rat and a mouse. All that we are doing here is to place as well, in the same multidimensional system of axes, the very much larger set of animals that have never existed. We are including those that could never have survived even if they had come into existence, as well as those that might have survived if they had existed but as a matter of fact never came into existence.

Movement from one point in the landscape to another is mutation, interpreted in its broadest sense to include large-scale changes in the genetic system as well as point mutations at loci within existing genetic systems. In principle, by a sufficiently contrived piece of genetic engineering – artificial mutation – it is possible to move from any point in the landscape to any other. There exists a recipe for transforming the genome of a human into the genome of a hippo or into the genome of any other animal, actual or conceivable. It would normally be a very large recipe, involving changes to many of the genes, deletion of many genes, duplication of many genes, and radical reorganizations of the genetic system. Nevertheless, the recipe is in principle discoverable, and obeying it can be represented as equivalent to taking a single giant leap from one point to another in our mathematical space. In practice, viable mutations are normally relatively small steps in the landscape: children are only slightly

*I find this image, which is modified from the venerable American population geneticist Sewall Wright, a helpful way to think about evolution. I first made use of it in *The Blind Watchmaker* and gave it two chapters in *Climbing Mount Improbable*, where I called it a 'museum' of all possible animals. Museum is superficially better than landscape because it is three-dimensional, although actually, of course, we are usually dealing with many more than three dimensions. Daniel Dennett's version, in *Darwin's Dangerous Idea*, is a library, the vividly named 'Library of Mendel'.

different from their parents even if, in principle, they could be as different as a hippo is from a human. Evolution consists of step-by-step trajectories through the genetic space, not large leaps. Evolution, in other words, is gradualistic. There is a general reason why this has to be so, a reason that I shall now develop.

Even without formal mathematical treatment, we can make some statistical statements about our landscape. First, in the landscape of all possible genetic combinations and the 'organisms' that they might generate, the proportion of viable organisms to nonviable organisms is very small. 'However many ways there may be of being alive, it is certain that there are vastly more ways of being dead.'[61] Second, taking any given starting point in the landscape, however many ways there may be of being slightly different, it is obvious that there are vastly more ways of being very different. The number of near neighbours in the landscape may be large, but it is dwarfed by the number of distant neighbours. As we consider hyperspheres of ever increasing size, the number of progressively more distant genetic neighbours that the spheres envelop mounts as a power function and rapidly becomes for practical purposes infinite.

The statistical nature of this argument points up an irony in the claim, frequently made by lay opponents of evolution, that the theory of evolution violates the Second Law of thermodynamics, the law of increasing entropy or chaos* within any closed system. The truth is opposite. If anything appeared to violate the law (nothing really does), it would be the *facts*†, not any particular explanation of those facts! The Darwinian explanation, indeed, is the only viable explanation we have for those facts that shows us how they could have come into being *without* violating the laws of physics. The law of increasing entropy is, in any case, subject to an interesting misunderstanding, which is worthy of a brief digression because it has helped to foster the mistaken claim that the idea of evolution violates the law.

The Second Law originated in the theory of heat engines,[62] but the form of it that is relevant to the evolutionary argument can be stated in more general statistical terms. Entropy was characterized

---

*Chaos here has its original and still colloquial meaning, not the technical meaning which it has recently acquired.
†About life's functional complexity or high 'information content'.

by the physicist Willard Gibbs as the 'mixed-upness' of a system. The law states that the total entropy of a system and its surroundings will not decrease. Left to itself, without work being contributed from outside, any closed system (life is not a closed system) will tend to become more mixed-up, less orderly. Homely analogies – or they may be more than analogies – abound. If there is not constant work being put in by a librarian, the orderly shelving of books in a library will suffer relentless degradation due to the inevitable if low probability that borrowers will return them to the wrong shelf. We have to import a hard-working librarian into the system from outside, who, Maxwell's-Demon-like, methodically and energetically restores order to the shelves.

The common error to which I referred is to personify the Second Law: to invest the universe with an inner urge or drive towards chaos; a positive striving towards an ultimate nirvana of perfect disorder. It is partly this error that has led people to accept the foolish notion that evolution is a mysterious exception to the law. The error can most simply be exposed by reference to the library analogy. When we say that an unattended library tends to approach chaos as time proceeds, we do not mean that any particular state of the shelves is being approached, as though the library were striving towards a goal from afar. Quite the contrary. The number of possible ways of shelving the N books in a library can be calculated, and for any nontrivial library it is a very, very large number indeed. Of these ways, only one, or a very few, would be recognized by us as a state of order. That is all there is to it. Far from there being any mystical urge towards disorder, it is just that there are vastly more ways of being recognized as disorderly than of being recognized as orderly. So, if a system wanders anywhere in the space of all possible arrangements, it is almost certain – unless special, librarian-like steps are taken – that we shall perceive the change as an increase in disorder. In the present context of evolutionary biology, the particular kind of order that is relevant is adaptation, the state of being equipped to survive and reproduce.

Returning to the general argument in favour of gradualism, to find viable life forms in the space of all possible forms is like searching for a modest number of needles in an extremely large haystack. The chance of happening to land on one of the needles if we take a large random mutational leap to another place in our multidimensional haystack is very small indeed. But one thing we can say is that the starting point of any mutational leap has to be a viable organism – one of the rare and precious needles in the haystack. This is because only organisms good enough to survive to reproductive age can have offspring of any kind, including mutant offspring. Finding a viable body-form by random mutation may be like finding a needle in a haystack, but given that you have already found one viable body-form, it is certain that you can hugely increase your chances of finding another viable one if you search in the immediate neighbourhood rather than more distantly.

The same goes for finding an improved body-form. As we consider mutational leaps of decreasing magnitude, the absolute number of destinations decreases but the proportion of destinations that are improvements increases. Fisher gave an elegantly simple argument to show that this increase tends towards 50 per cent for mutational changes of very small magnitude.* His argument seems inescapable for any single dimension of variation considered on its own. Whether his precise conclusion (50 per cent) generalizes to the multidimensional case I shall not discuss, but the direction of the argument is surely indisputable. The larger the leap through genetic space, the lower is the probability that the resulting change will be viable, let alone an improvement. Gradualistic, step-by-step walking in the immediate vicinity of already discovered needles in the haystack seems to be the only way to find other and better needles. Adaptive evolution must in general be a crawl through genetic space, not a series of leaps.

But are there any special occasions when macromutations are incorporated into evolution? Macromutations certainly occur in the laboratory.† Our theoretical considerations say only that *viable* macromutations should be exceedingly rare in comparison with viable micromutations. But even if the occasions when

*He used the analogy of perfecting the focus of a microscope. A very small movement of the objective lens has a 50 per cent chance of being in the right direction (which will improve the focus). A large movement is bound to make things worse (even if it was in the right direction, it will overshoot).
†Macromutations, or saltations, are mutations of large magnitude. A famous example in fruit flies is *antennapedia*. Mutant flies grow a leg where an antenna should be.

major saltations are viable and incorporated into evolution are exceedingly rare, even if they have occurred only once or twice in the whole history of a lineage from Precambrian to present, that is enough to transform the entire course of evolution. I find it plausible, for instance, that the invention of segmentation occurred in a single macromutational leap, once during the history of our own vertebrate ancestors and again once in the ancestry of arthropods and annelids. Once this had happened, in each of these two lineages, it changed the entire climate in which ordinary cumulative selection of micromutations went on. It must have resembled, indeed, a sudden catastrophic change in the external climate. Just as a lineage can, after appalling loss of life, recover and adapt to a catastrophic change in the external climate, so a lineage might, by subsequent micromutational selection, adapt to the catastrophe of a macromutation as large as the first segmentation.

In the landscape of all possible animals, our segmentation example might look like this. A wild macromutational leap from a perfectly viable parent lands in a remote part of the haystack, far from any needle of viability. The first segmented animal is born: a freak; a monster none of whose detailed bodily features equip it to survive its new, segmented architecture. It should die. But by chance the leap in genetic space has coincided with a leap in geographical space. The segmented monster finds itself in a virgin part of the world where the living is easy and competition is light. What can happen when any ordinary animal finds itself in a strange place, a new continent, say, is that, although ill-adapted to the new conditions, it survives by the skin of its teeth. In the competition vacuum, its descendants survive for enough generations to adapt, by normal, cumulative natural selection of micromutations, to the alien conditions. So it might have been with our segmented monster. It survived by the skin of its teeth, and its descendants adapted, by ordinary micromutational cumulative selection, to the radically new conditions imposed by the macromutation. Though the macromutational leap landed far from any needle in the haystack, the competition vacuum enabled the monster's descendants subsequently to inch their way towards the nearest needle. As it turned out, when all the compensating evolution at other genetic loci had been completed, the body plan represented by that nearest needle eventually emerged as superior to the ancestral unsegmented body plan. The new local optimum, into whose vicinity the lineage wildly leapt, eventually turned out superior to the local optimum on which it had previously been trapped.

This is the kind of speculation in which we should indulge only as a last resort. The argument stands that only gradualistic, inch-by-inch walking through the genetic landscape is compatible with the sort of cumulative evolution that can build up complex and detailed adaptation. Even if segmentation, in our example, ended up as a superior body form, it began as a catastrophe that had to be weathered, just like a climatic or volcanic catastrophe in the external environment. It was gradualistic, cumulative selection that engineered the step-by-step recovery from the segmentation catastrophe, just as it engineers recoveries from external climatic catastrophes. Segmentation, according to the speculation I have just given, survived not because natural selection favoured it but because natural selection found compensatory ways of survival *in spite of it*. The fact that advantages in the segmented body plan eventually emerged is an irrelevant bonus. The segmented body plan was incorporated into evolution, but it may never have been favoured by natural selection.

But in any case gradualism is only a part of core Darwinism. A belief in the ubiquity of gradualistic evolution does not necessarily commit us to Darwinian natural selection as the steering mechanism guiding the search through genetic space. It is highly probable that Motoo Kimura is right to insist that most of the evolutionary steps taken through genetic space are unsteered steps. To a large extent the trajectory of small, gradualistic steps actually taken may constitute a random walk rather than a walk guided by selection. But this is irrelevant if – for the reasons given above – our concern is with adaptive evolution as opposed to evolutionary change *per se*. Kimura himself rightly insists* that his 'neutral theory is not antagonistic to

---

*'Insists' may be putting it a bit strongly. Now that Professor Kimura is dead, the rather endearing story told by John Maynard Smith can be included. It is true that Kimura's book includes the statement that natural selection must be involved in adaptive evolution but, according to Maynard Smith, Kimura could not bear to write the sentence himself and he asked his friend, the distinguished American geneticist James Crow, to write it for him. The book is M. Kimura, *The Neutral Theory of Molecular Evolution* (Cambridge, Cambridge University Press, 1983).

the cherished view that evolution of form and function is guided by Darwinian selection'. Further,

> the theory does not deny the role of natural selection in determining the course of adaptive evolution, but it assumes that only a minute fraction of DNA changes in evolution are adaptive in nature, while the great majority of phenotypically silent molecular substitutions exert no significant influence on survival and reproduction and drift randomly through the species.

The facts of adaptation compel us to the conclusion that evolutionary trajectories are not all random. There has to be some nonrandom guidance towards adaptive solutions because non-random is what adaptive solutions precisely are. Neither random walk nor random saltation can do the trick on its own. But does the guiding mechanism necessarily have to be the Darwinian one of nonrandom survival of random spontaneous variation? The obvious alternative class of theory postulates some form of nonrandom, i.e. directed, *variation*.

Nonrandom, in this context, means directed towards adaptation. It does not mean causeless. Mutations are, of course, caused by physical events, for instance, cosmic ray bombardment. When we call them random, we mean only that they are random with respect to adaptive improvement.[63] It could be said, therefore, that, as a matter of logic, some kind of theory of directed variation is the only alternative to natural selection as an explanation for adaptation. Obviously, combinations of the two kinds of theory are possible.

The theory nowadays attributed to Lamarck is typical of a theory of directed variation. It is normally expressed as two main principles. First, organisms improve during their own lifetime by means of the principle of use and disuse; muscles that are exercised as the animal strives for a particular kind of food enlarge, for instance, and the animal is consequently better equipped to procure that food in the future. Second, acquired characteristics – in this case acquired improvements due to use – are inherited so, as the generations go by, the lineage improves. Arguments offered against Lamarckian theories are usually factual. Acquired charac-teristics are not, as a matter of fact, inherited. The implication, often made explicit, is that if only they were inherited, Lamarckism would be a tenable theory of evolution. Ernst Mayr,[64] for instance, wrote,

> Accepting his premises, Lamarck's theory was as legitimate a theory of adaptation as that of Darwin. Unfortunately, these premises turned out to be invalid.

Francis Crick[65] showed an awareness of the possibility that general *a priori* arguments might be given, when he wrote,

> As far as I know, no one has given general theoretical reasons why such a mechanism must be less efficient than natural selection.

I have since offered two such reasons, following an argument that the inheritance of acquired characteristics is *in principle* incompatible with embryology as we know it.[66]

First, acquired improvements could in principle be inherited only if embryology were *preformationistic* rather than *epigenetic*. Preformationistic embryology is blueprint embryology. The alternative is recipe, or computer-program, embryology. The important point about blueprint embryology is that it is reversible. If you have a house, you can, by following simple rules, reconstruct its blueprint. But if you have a cake, there is no set of simple rules that enables you to reconstruct its recipe. All living things on this planet grow by recipe embryology, not blueprint embryology. The rules of development work only in the forward direction, like the rules in a recipe or computer program. You cannot, by inspecting an animal, reconstruct its genes. Acquired characteristics are attributes of the animal. In order for them to be inherited, the animal would have to be scanned and its attributes reverse-transcribed into the genes. There may be planets whose animals develop by blueprint embryology. If so, acquired characteristics might there be inherited. This argument says that if you want to find a Lamarckian form of life, don't bother to look on any planet whose life forms develop by epigenesis rather than preformationism. I have an intuitive hunch that there may be a general, *a priori* argument against preformationistic, blueprint embryology, but I have not developed it yet.

Second, most acquired characteristics are not improvements.

There is no general reason why they should be, and use and disuse does not really help here. Indeed, by analogy with wear and tear on machines, we might expect use and disuse to be positively counterproductive. If acquired characteristics were indiscriminately inherited, organisms would be walking museums of ancestral decrepitude, pock-marked from ancestral plagues, limping relics of ancestral misfortune. How is the organism supposed to 'know' how to respond to the environment in such a way as to improve itself? If there is a minority of acquired characteristics that are improvements, the organism would have to have some way of selecting these to pass on to the next generation, avoiding the much more numerous acquired characteristics that are deleterious. Selecting, here, really means that some form of Darwinian process must be smuggled in. Lamarckism cannot work unless it has a Darwinian underpinning.

Third, even if there were some means of choosing which acquired characteristics should be inherited, which discarded at the current generation, the principle of use and disuse is not powerful enough to fashion adaptations as subtle and intricate as we know them to be. A human eye, for instance, works well because of countless pernickety adjustments of detail. Natural selection can fine-tune these adjustments because any improvement, however slight and however deeply buried in internal architecture, can have a direct effect upon survival and reproduction. The principle of use and disuse, on the other hand, is in principle incapable of such fine-tuning. This is because it relies upon the coarse and crude rule that the more an animal uses a bit of itself, the bigger that bit ought to be. Such a rule might tune the blacksmith's arms to his trade, or the giraffe's neck to the tall trees. But it could hardly be responsible for improving the lucidity of a lens or the reaction time of an iris diaphragm. The correlation between use and size is too loose to be responsible for fine-grained adaptation.

I shall refer to these three arguments as the 'Universal Darwinism' arguments. I am confident that they are arguments of the kind that Crick was calling for, although whether he or anyone else accepts these three particular arguments is another matter. If they are correct, the case for Darwinism, in its most general form, is enormously strengthened.

I suspect that other armchair arguments about the nature of life all over the universe, more powerful and watertight than mine, are waiting to be discovered by those better equipped than I am. But I cannot forget that Darwin's own triumph, for all that it *could* have been launched from any armchair in the universe, was in fact the spin-off of a five-year circumnavigation of this particular planet.

The 'Information Challenge'[67]

In September 1997, I allowed an Australian film crew into my house in Oxford without realizing that their purpose was creationist propaganda. In the course of a suspiciously amateurish interview, they issued a truculent challenge to me to 'give an example of a genetic mutation or an evolutionary process which can be seen to increase the information in the genome'. It is the kind of question only a creationist would ask in that way, and it was at this point I tumbled to the fact that I had been duped into granting an interview to creationists – a thing I normally don't do, for good reasons.* In my anger I refused to discuss the question further, and told them to stop the camera. However, I eventually withdrew my peremptory termination of the interview, because they pleaded with me that they had come all the way from Australia specifically in order to interview me. Even if this was a considerable exaggeration, it seemed, on reflection, ungenerous to tear up the legal release form and throw them out. I therefore relented.

My generosity was rewarded in a fashion that anyone familiar with fundamentalist tactics might have predicted. When I eventually saw the film a year later,† I found that it had been edited to give the false impression that I was *incapable* of answering the question about information content.‡ In fairness, this may not have been quite as intentionally deceitful as it sounds.

*See 'Unfinished Correspondence with a Darwinian Heavyweight' (pp. 256–62).
†The producers never deigned to send me a copy: I completely forgot about it until an American colleague called it to my attention.
‡See Barry Williams, 'Creationist deception exposed', *the Skeptic* **18** (1998), 3, pp. 7–10, for an account of how my long pause (trying to decide whether to throw them out) was made to look like hesitant inability to answer the question, followed by an apparently evasive answer to a completely different question.

You have to understand that these people really *believe* that their question *cannot* be answered! Pathetic as it sounds, their entire journey from Australia seems to have been a quest to film an evolutionist failing to answer it.

With hindsight – given that I had been suckered into admitting them into my house in the first place – it might have been wiser simply to answer the question. But I like to be understood whenever I open my mouth – I have a horror of blinding people with science – and this was not a question that could be answered in a soundbite. First you have to explain the technical meaning of 'information'. Then the relevance to evolution, too, is complicated – not really difficult but it takes time. Rather than engage in further recriminations and disputes about exactly what happened at the time of the interview, I shall try to redress the matter now in constructive fashion by answering the original question, the 'Information Challenge', at adequate length – the sort of length you can achieve in a proper article.

The technical definition of 'information' was introduced by the American engineer Claude Shannon in 1948. An employee of the Bell Telephone Company, Shannon was concerned to measure information as an economic commodity. It is costly to send messages along a telephone line. Much of what passes in a message is not information: it is *redundant*. You could save money by recoding the message to remove the redundancy. Redundancy was a second technical term introduced by Shannon, as the inverse of information. Both definitions are mathematical, but we can convey Shannon's intuitive meaning in words.* Redundancy is any part of a message that is not informative, either because the recipient already knows it (is not surprised by it) or because it

---

*It is important not to blame Shannon for my verbal and intuitive way of expressing what I think of as the essence of his idea. Mathematical readers should go straight to the original, C. Shannon and W. Weaver, *The Mathematical Theory of Communication* (University of Illinois Press, 1949). Claude Shannon, by the way, had an imaginative sense of humour. He once built a box with a single switch on the outside. If you threw the switch, the lid of the box slowly opened, a mechanical hand appeared, reached down and switched off the box. It then put itself away and the lid closed. As Arthur C. Clarke said, 'There is something unspeakably sinister about a machine that does nothing – absolutely nothing – except switch itself off.'

duplicates other parts of the message. In the sentence 'Rover is a poodle dog', the word 'dog' is redundant because 'poodle' already tells us that Rover is a dog. An economical telegram would omit it, thereby increasing the informative proportion of the message. 'Arr JFK Fri pm pls mt BA Cncrd flt' carries the same information as the much longer, but more redundant, 'I'll be arriving at John F Kennedy airport on Friday evening; please meet the British Airways Concorde flight'. Obviously the brief, telegraphic message is cheaper to send (although the recipient may have to work harder to decipher it – redundancy has its virtues if we forget economics). Shannon wanted to find a mathematical way to capture the idea that any message could be broken into the *information* (which is worth paying for), the *redundancy* (which can, with economic advantage, be deleted from the message because, in effect, it can be reconstructed by the recipient) and the *noise* (which is just random rubbish).

'It rained in Oxford every day this week' carries relatively little information, because the receiver is not surprised by it. On the other hand, 'It rained in the Sahara desert every day this week' would be a message with high information content, well worth paying extra to send. Shannon wanted to capture this sense of information content as 'surprise value'. It is related to the other sense – 'that which is not duplicated in other parts of the message' – because repetitions lose their power to surprise. Note that Shannon's definition of the quantity of information is independent of whether it is true. The measure he came up with was ingenious and intuitively satisfying. Let's estimate, he suggested, the receiver's ignorance or uncertainty *before* receiving the message, and then compare it with the receiver's remaining ignorance *after* receiving the message. The quantity of ignorance-reduction is the information content. Shannon's unit of information is the *bit*, short for 'binary digit'. One bit is defined as the amount of information needed to halve the receiver's prior uncertainty, however great that prior uncertainty was (mathematical readers will notice that the bit is, therefore, a logarithmic measure).

In practice, you first have to find a way of measuring the prior uncertainty – that which is reduced by the information when it comes. For particular kinds of simple message, this is easily done in terms of probabilities. An expectant father watches the birth of his child through a window. He can't see any details, so a nurse has agreed to hold up a pink card if it is a girl, blue for a boy. How much information is conveyed when, say, the nurse flourishes the pink card to the delighted father? The answer is one *bit* – the prior uncertainty is halved. The father knows that a baby of some kind has been born, so his uncertainty amounts to just two possibilities – boy and girl – and they are (for purposes of this discussion) equiprobable. The pink card *halves* the father's prior uncertainty from two possibilities to one (girl). If there'd been no pink card but a doctor walked out of the room, shook the father's hand and said, 'Congratulations old chap, I'm delighted to be the first to tell you that you have a daughter', the information conveyed by the 17-word message would still be only one bit.

Computer information is held in a sequence of noughts and ones. There are only two possibilities, so each 0 or 1 can hold one bit. The memory capacity of a computer, or the storage capacity of a disk or tape, is often measured in bits, and this is the total number of 0s or 1s that it can hold. For some purposes, more convenient units of measurement are the byte (8 bits), the kilobyte (1000 bytes), the megabyte (a million bytes) or the gigabyte (1000 million bytes).* Notice that these figures refer to the total available capacity. This is the maximum quantity of information that the device is capable of storing. The actual amount of information stored is something else. The capacity of my hard disk happens to be 4.2 gigabytes. Of this, about 1.4 gigabytes are actually being used to store data at present. But even this is not the true information content of the disk in Shannon's sense. The true information content is smaller, because the

*These round figures are all decimal approximations. In the world of computers, the standard metric prefixes, 'kilo', 'giga' etc. are borrowed for the nearest convenient power of 2. Thus a kilobyte is not 1000 bytes but 2^{10} or 1024 bytes; a megabyte is not a million bytes but 2^{20} or 1,048,576 bytes. If we had evolved with 8 fingers or 16, instead of 10, the computer might have been invented a century earlier. Theoretically, we could now decide to teach all children octal instead of decimal arithmetic. I'd love to give it a go, but realistically I recognize that the immense short-term costs of the transition would outweigh the undoubted long-term benefits of the change. For a start, we'd all have to learn our multiplication tables again from scratch.

information could be more economically stored. You can get some idea of the true information content by using one of those ingenious compression programs like 'Stuffit'. Stuffit looks for redundancy in the sequence of 0s and 1s, and removes a hefty proportion of it by recoding – stripping out internal predictability. Maximum information content would be achieved (probably never in practice) only if every 1 or 0 surprised us equally. Before data is transmitted in bulk around the internet, it is routinely compressed to reduce redundancy.*

That's good economics. But on the other hand it is also a good idea to keep some redundancy in messages, to help correct errors. In a message that is totally free of redundancy, after there's been an error there is no means of reconstructing what was intended. Computer codes often incorporate deliberately redundant 'parity bits' to aid in error detection. DNA, too, has various error-correcting procedures which depend upon redundancy. When I come on to talk of genomes, I'll return to the three-way distinction between total information capacity, information capacity actually used, and true information content.

It was Shannon's insight that information of any kind, no matter what it means, no matter whether it is true or false, and no matter by what physical medium it is carried, can be measured in bits, and is translatable into any other medium of information. The great biologist J. B. S. Haldane used Shannon's theory to compute the number of bits of information conveyed by a worker bee to her hivemates when she 'dances' the location of a food source (about 3 bits to tell about the direction of the food and another 3 bits for the distance of the food). In the same units, I recently calculated that I'd need to set aside 120 megabits of laptop computer memory to store the triumphal opening chords of Richard Strauss's *Also Sprach Zarathustra* (the '2001 theme'),

---

*A powerful application of this aspect of information theory is Horace Barlow's idea that sensory systems are wired up to remove massive amounts of redundancy before passing their messages on to the brain. One way they do this is by signalling *change* in the world (what mathematicians would call differentiating) rather than continuously reporting the current state of the world (which is highly redundant because it doesn't fluctuate rapidly and randomly). I discussed Barlow's idea in *Unweaving the Rainbow* (London, Penguin, 1998; Boston, Houghton Mifflin, 1998), pp. 257–66.

which I wanted to play in the middle of a lecture about evolution. Shannon's economics enable you to calculate how much modem time it'll cost you to email the complete text of a book to a publisher in another land. Fifty years after Shannon, the idea of information as a commodity, as measurable and interconvertible as money or energy, has come into its own.

DNA carries information in a very computer-like way, and we can measure the genome's capacity in bits too, if we wish. DNA doesn't use a binary code, but a quaternary one. Whereas the unit of information in the computer is a 1 or a 0, the unit in DNA can be T, A, C or G. If I tell you that a particular location in a DNA sequence is a T, how much information is conveyed from me to you? Begin by measuring the prior uncertainty. How many possibilities are open before the message 'T' arrives? Four. How many possibilities remain after it has arrived? One. So you might think the information transferred is four bits, but actually it is two. Here's why (assuming that the four letters are equally probable, like the four suits in a pack of cards). Remember that Shannon's metric is concerned with the most *economical* way of conveying the message. Think of it as the number of yes/no questions that you'd have to ask in order to narrow down to certainty, from an initial uncertainty of four possibilities, assuming that you planned your questions in the most *economical* way. 'Is the mystery letter before D in the alphabet?'* No. That narrows it down to T or G, and now we need only one more question to clinch it. So, by this method of measuring, each 'letter' of the DNA has an information capacity of 2 bits.

Whenever prior uncertainty of recipient can be expressed as a number of equiprobable alternatives N, the information content of a message which narrows those alternatives down to one is $\log_2 N$ (the power to which 2 must be raised in order to yield the number of alternatives N). If you pick a card, any card, from a normal pack, a statement of the identity of the card carries $\log_2 52$,

---

*A chemist would more naturally ask, 'Is it a pyrimidine?', but that sends the wrong signal for my purposes. It is only *incidentally* true that the four letters of the DNA alphabet fall naturally into two chemical families, purines and pyrimidines.

or 5.7 bits of information. In other words, given a large number of guessing games, it would take 5.7 yes/no questions on average to guess the card, provided the questions are asked in the most economical way. The first two questions might establish the suit (Is it red? Is it a diamond?); the remaining three or four questions would successively divide and conquer the suit (Is it a 7 or higher? etc.), finally homing in on the chosen card. When the prior uncertainty is some mixture of alternatives that are not equiprobable, Shannon's formula becomes a slightly more elaborate weighted average, but it is essentially similar. By the way, Shannon's weighted average is the same formula as physicists have used, since the nineteenth century, for entropy. The point has interesting implications but I shall not pursue them here.*

That's enough background on information theory. It is a theory which has long held a fascination for me, and I have used it in several of my research papers over the years. Let's now think how we might use it to ask whether the information content of genomes increases in evolution. First, recall the three-way distinction between total information capacity, the capacity that is actually used, and the true information content when stored in the most economical way possible. The total information capacity of the human genome is measured in gigabits. That of the common gut bacterium *Escherichia coli* is measured in megabits. We, like all other animals, are descended from an ancestor which, were it available for our study today, we'd classify as a bacterium. So during the billions of years of evolution since that ancestor lived, the information capacity of our genome has gone up perhaps three orders of magnitude (powers of ten) – about a thousandfold. This is satisfyingly plausible and comforting to human dignity.

Should human dignity feel wounded, then, by the fact that the crested newt, *Triturus cristatus*, has a genome capacity estimated at 40 gigabits, an order of magnitude larger than the human genome? No, because, in any case, most of the capacity of the genome of any animal is not used to store useful information. There are many nonfunctional pseudogenes (see below) and lots

*Ecologists also use the formula as an index of diversity.

of repetitive nonsense, useful for forensic detectives but not translated into protein in the living cells. The crested newt has a bigger 'hard disk' than we have, but since the great bulk of both our hard disks is unused, we needn't feel insulted. Related species of newt have much smaller genomes. Why the Creator should have played fast and loose with the genome sizes of newts in such a capricious way is a problem that creationists might like to ponder. From an evolutionary point of view the explanation is simple.*

Evidently the total information capacity of genomes is very variable across the living kingdoms, and it must have changed greatly in evolution, presumably in both directions. Losses of genetic material are called deletions. New genes arise through various kinds of duplication. This is well illustrated by haemoglobin, the complex protein molecule that transports oxygen in the blood.

Human adult haemoglobin is actually a composite of four protein chains called globins, knotted around each other. Their detailed sequences show that the four globin chains are closely related to each other, but they are not identical. Two of them are called alpha globins (each a chain of 141 amino acids), and two are beta globins (each a chain of 146 amino acids). The genes coding for the alpha globins are on chromosome 11; those coding for the beta globins are on chromosome 16. On each of these chromosomes, there is a cluster of globin genes in a row, interspersed with some junk DNA. The alpha cluster, on chromosome 11, contains seven globin genes. Four of these are pseudogenes, versions of alpha disabled by faults in their sequence and not translated into proteins. Two are true alpha globins, used in the adult. The final one is called zeta and is used only in embryos. Similarly the beta cluster, on chromosome 16, has six genes, some of which are disabled, and one of which is used only in the embryo. Adult haemoglobin, as we've seen, contains two alpha and two beta chains.

Never mind all this complexity. Here's the fascinating point.

*My suggestion (*The Selfish Gene*, 1976) that surplus DNA is parasitic was later taken up and developed by others under the catch-phrase 'Selfish DNA'. See *The Selfish Gene*, 2nd edn (Oxford University Press, 1989), pp. 44–5 and 275.

Careful letter-by-letter analysis shows that these different kinds of globin genes are literally cousins of each other, literally members of a family. But these distant cousins still coexist inside our own genome, and that of all vertebrates. On the scale of whole organisms, all vertebrates are our cousins too. The tree of vertebrate evolution is the family tree we are all familiar with, its branch-points representing speciation events – the splitting of species into pairs of daughter species. But there is another family tree occupying the same timescale, whose branches represent not speciation events but gene duplication events within genomes.

The dozen or so different globins inside you are descended from an ancient globin gene which, in a remote ancestor who lived about half a billion years ago, duplicated, after which both copies stayed in the genome. There were then two copies of it, in different parts of the genome of all descendant animals. One copy was destined to give rise to the alpha cluster (on what would eventually become chromosome 11 in our genome), the other to the beta cluster (on chromosome 16). As the aeons passed, there were further duplications (and doubtless some deletions as well). Around 400 million years ago the ancestral alpha gene duplicated again, but this time the two copies remained near neighbours of each other, in a cluster on the same chromosome. One of them was destined to become the zeta used by embryos, the other became the alpha globin genes used by adult humans (other branches gave rise to the nonfunctional pseudogenes I mentioned). It was a similar story along the beta branch of the family, but with duplications at other moments in geological history.

Now here's an equally fascinating point. Given that the split between the alpha cluster and the beta cluster took place 500 million years ago, it will of course not be just our human genomes that show the split – that is, possess alpha genes in a different part of the genome from beta genes. We should see the same within-genome split if we look at any other mammals, at birds, reptiles, amphibians and bony fish, for our common ancestor with all of them lived less than 500 million years ago. Wherever it has been investigated, this expectation has proved correct. Our greatest hope of finding a vertebrate that does not share with us the ancient alpha/beta split would be a jawless fish like a lamprey, for they are our most remote cousins among surviving vertebrates; they are the only surviving vertebrates whose common ancestor with the rest of the vertebrates is sufficiently ancient that it could have predated the alpha/beta split. Sure enough, these jawless fishes are the only known vertebrates that lack the alpha/beta divide.

Gene duplication, within the genome, has a similar historic impact to species duplication ('speciation') in phylogeny. It is responsible for gene diversity, in the same way as speciation is responsible for phyletic diversity. Beginning with a single universal ancestor, the magnificent diversity of life has come about through a series of branchings of new species, which eventually gave rise to the major branches of the living kingdoms and the hundreds of millions of separate species that have graced the Earth. A similar series of branchings, but this time within genomes – gene duplications – has spawned the large and diverse population of clusters of genes that constitutes the modern genome.

The story of the globins is just one among many. Gene duplications and deletions have occurred from time to time throughout genomes. It is by these, and similar means, that genome sizes can increase in evolution. But remember the distinction between the total capacity of the whole genome, and the capacity of the portion that is actually used. Recall that not all the globin genes are used. Some of them, like theta in the alpha cluster of globin genes, are pseudogenes, recognizably kin to functional genes in the same genomes, but never actually translated into the action language of protein. What is true of globins is true of most other genes. Genomes are littered with nonfunctional pseudogenes, faulty duplicates of functional genes that do nothing, while their functional cousins (the word doesn't even need scare quotes) get on with their business in a different part of the same genome. And there's lots more DNA that doesn't even deserve the name pseudogene. It too is derived by duplication, but not duplication of functional genes. It consists of multiple copies of junk, 'tandem repeats', and other nonsense which may be useful for forensic detectives but which doesn't seem to be used in the body itself. Once again, creationists might spend some earnest time speculating on why the Creator should bother to litter genomes with untranslated pseudogenes and junk tandem repeat DNA.

Can we measure the information capacity of that portion of the genome which is actually used? We can at least estimate it. In the case of the human genome it is about 2 per cent – considerably less than the proportion of my hard disk that I have used since I bought it. Presumably the equivalent figure for the crested newt is even smaller, but I don't know if it has been measured. In any case, we mustn't run away with a chauvinistic idea that the human genome somehow ought to have the largest DNA database because we are so wonderful. The great evolutionary biologist George C. Williams has pointed out that animals with complicated life cycles need to code for the development of all stages in the life cycle, but they only have one genome with which to do so. A butterfly's genome has to hold the complete information needed for building a caterpillar as well as a butterfly. A sheep liver fluke has six distinct stages in its life cycle, each specialized for a different way of life. We shouldn't feel too insulted if liver flukes turned out to have bigger genomes than we have (actually they don't).

Remember, too, that even the total capacity of genome that is actually used is still not the same thing as the true information content in Shannon's sense. The true information content is what's left when the redundancy has been compressed out of the message, by the theoretical equivalent of Stuffit. There are even some viruses that seem to use a kind of Stuffit-like compression. They make use of the fact that the RNA (not DNA in these viruses, as it happens) code is read in triplets. There is a 'frame' which moves along the RNA sequence, reading off three letters at a time. Obviously, under normal conditions, if the frame starts reading in the wrong place (as in a so-called frame-shift mutation), it makes total nonsense: the 'triplets' that it reads are out of step with the meaningful ones. But these splendid viruses actually exploit frame-shifted reading. They get two messages for the price of one, by having a completely different message embedded in the very same series of letters when read frame-shifted. In principle you could even get three messages for the price of one, but I don't know of any examples.

It is one thing to estimate the total information capacity of a genome, and the amount of the genome that is actually used, but it's harder to estimate its true information content in the Shannon sense. The best we can do is probably to forget about the genome itself and look at its product, the 'phenotype', the working body of the animal or plant itself. In 1951, J. W. S. Pringle, who later became my Professor at Oxford, suggested using a Shannon-type information measure to estimate 'complexity'. Pringle wanted to express complexity mathematically in *bits*, but I have long found the following verbal form helpful in explaining his idea.

We have an intuitive sense that a lobster, say, is more complex (more 'advanced', some might even say more 'highly evolved') than another animal, perhaps a millipede. Can we *measure* something in order to confirm or deny our intuition? Without literally turning it into bits, we can make an approximate estimation of the information contents of the two bodies as follows. Imagine writing a book describing the lobster. Now write another book describing the millipede down to the same level of detail. Divide the word-count in one book by the word-count in the other, and you have an approximate estimate of the relative information content of lobster and millipede. It is important to specify that both books describe their respective animals 'down to the same level of detail'. Obviously, if we describe the millipede down to cellular detail, but stick to gross anatomical features in the case of the lobster, the millipede would come out ahead.

But if we do the test fairly, I'll bet the lobster book would come out longer than the millipede book. It's a simple plausibility argument, as follows. Both animals are made up of segments – modules of bodily architecture that are fundamentally similar to each other, arranged fore-and-aft like the trucks of a train. The millipede's segments are mostly identical to each other. The lobster's segments, though following the same basic plan (each with a nervous ganglion, a pair of appendages, and so on) are mostly different from each other. The millipede book would consist of one chapter describing a typical segment, followed by the phrase 'Repeat *N* times', where *N* is the number of segments. The lobster book would need a different chapter for each segment. This isn't quite fair on the millipede, whose front and rear end segments are a bit different from the rest. But I'd still bet that, if anyone bothered to do the experiment, the estimate of lobster information content would come out substantially greater than the estimate of millipede information content.

It's not of direct evolutionary interest to compare a lobster with a millipede in this way, because nobody thinks lobsters evolved from millipedes. Obviously no modern animal evolved from any other modern animal. Instead, any pair of modern animals had a last common ancestor which lived at some (in principle) discoverable moment in geological history. Almost all of evolution happened way back in the past, which makes it hard to study details. But we can use the 'length of book' thought-experiment to agree upon what it would *mean* to ask the question whether information content increases over evolution, if only we had ancestral animals to look at.

The answer in practice is complicated and controversial, all bound up with a vigorous debate over whether evolution is, in general, progressive. I am one of those associated with a limited form of yes answer. My colleague Stephen Jay Gould tends towards a no answer.* I don't think anybody would deny that, by any method of measuring – whether bodily information content, total information capacity of genome, capacity of genome actually used, or true ('Stuffit compressed') information content of genome – there has been a broad overall trend towards increased information content during the course of human evolution from our remote bacterial ancestors. People might disagree, however, over two important questions: first, whether such a trend is to be found in all, or a majority of evolutionary lineages (for example, parasite evolution often shows a trend towards decreasing bodily complexity, because parasites are better off being simple); second, whether, even in lineages where there is a clear overall trend over the very long term, it is bucked by so many reversals and re-reversals in the shorter term as to undermine the very idea of progress. This is not the place to resolve this interesting controversy. There are distinguished biologists with good arguments on both sides.

*See 'Human Chauvinism and Evolutionary Progress' (pp. 242–55).

Supporters of 'intelligent design' guiding evolution, by the way, should be deeply committed to the view that information content increases during evolution. Even if the information comes from God, perhaps *especially* if it does, it should surely increase, and the increase should presumably show itself in the genome.

Perhaps the main lesson we should learn from Pringle is that the information content of a biological system is another name for its complexity. Therefore the creationist challenge with which we began is tantamount to the standard challenge to explain how biological complexity can evolve from simpler antecedents, one that I have devoted three books to answering, and I do not propose to repeat their contents here. The 'information challenge' turns out to be none other than our old friend: 'How could something as complex as an eye evolve?' It is just dressed up in fancy mathematical language – perhaps in an attempt to bamboozle. Or perhaps those who ask it have already bamboozled themselves, and don't realize that it is the same old – and thoroughly answered – question.

Let me turn, finally, to another way of looking at whether the information content of genomes increases in evolution. We now switch from the broad sweep of evolutionary history to the minutiae of natural selection. Natural selection itself, when you think about it, is a narrowing down from a wide initial field of possible alternatives, to the narrower field of the alternatives actually chosen. Random genetic error (mutation), sexual recombination and migratory mixing all provide a wide field of genetic variation: the available alternatives. Mutation is not an increase in true information content, rather the reverse, for mutation, in the Shannon analogy, contributes to increasing the prior uncertainty. But now we come to natural selection, which reduces the 'prior uncertainty' and therefore, in Shannon's sense, contributes information to the gene pool. In every generation, natural selection removes the less successful genes from the gene pool, so the remaining gene pool is a narrower subset. The narrowing is nonrandom, in the direction of improvement, where improvement is defined, in the Darwinian way, as improvement in fitness to survive and reproduce. Of course the total range of variation is topped up again in every generation by new mutation and other kinds of variation. But it still remains true that natural selection is a narrowing down from an initially wider field of possibilities, including mostly unsuccessful ones, to a narrower field of successful ones. This is analogous to the definition of information with which we began: information is what enables the narrowing down from prior uncertainty (the initial range of possibilities) to later certainty (the 'successful' choice among the prior probabilities). According to this analogy, natural selection is *by definition* a process whereby information is fed into the gene pool of the next generation.

If natural selection feeds information into gene pools, what is the information *about*? It is about how to survive. Strictly, it is about how to survive and reproduce, in the conditions that prevailed when previous generations were alive. To the extent that present day conditions are different from ancestral conditions, the ancestral genetic advice will be wrong. In extreme cases, the species may then go extinct. To the extent that conditions for the present generation are not too different from conditions for past generations, the information fed into present-day genomes from past generations is *helpful* information. Information from the ancestral past can be seen as a manual for surviving in the present: a family bible of ancestral 'advice' on how to survive today. We need only a little poetic licence to say that the information fed into modern genomes by natural selection is actually information about ancient environments in which ancestors survived.

This idea of information fed from ancestral generations into descendant gene pools is one of the themes of my book *Unweaving the Rainbow*. It takes a whole chapter, 'The Genetic Book of the Dead', to develop the notion, so I won't repeat it here except to say two things. First, it is the gene pool of the species as a whole, not the genome of any particular individual, which is best seen as the recipient of the ancestral information about how to survive. The genomes of particular individuals are random samples of the current gene pool, randomized by sexual recombination. Second, we are privileged to 'intercept' the information if we wish, and 'read' an animal's body, or even its genes, as a coded description of ancestral worlds. To quote from *Unweaving the Rainbow*:

And isn't it an arresting thought? We are digital archives of the African Pliocene, even of Devonian seas; walking repositories of wisdom out of the old days. You could spend a lifetime reading in this ancient library and die unsated by the wonder of it.

Genes Aren't Us[68]

The bogey of genetic determinism needs to be laid to rest. The discovery of a so-called 'gay gene' is as good an opportunity as we'll get to lay it.

The facts are quickly stated. In the magazine *Science*[69], a team of researchers from the National Institutes of Health, in Bethesda, Maryland, reported the following pattern. Homosexual males are more likely than you'd expect by chance to have homosexual brothers. Revealingly, they are also more likely than you'd expect by chance to have homosexual maternal uncles and homosexual cousins on the mother's side, but not on the father's side. This pattern raises the immediate suspicion that at least one gene causing homosexuality in males is carried on the X chromosome.*

The Bethesda team went further. Modern technology made it possible for them to search for particular marker strings in the DNA code itself. In one region, called Xq28, near the tip of the X chromosome, they found five identical markers shared by a suggestively high percentage of homosexual brothers. These facts combine elegantly with one another to confirm earlier evidence of a hereditary component to male homosexuality.

So what? Are sociology's foundations trembling? Should theologians be wringing their hands with concern, and lawyers rubbing theirs with anticipation? Does this finding tell us anything new about 'blame' or 'responsibility'? Does it add anything, one way or the other, to arguments about whether homosexuality is a condition that could, or should, be 'cured'? Should it make individual homosexuals more or less proud, or ashamed, of their

---

*Because males have only one X chromosome, which they necessarily get from their mother. Females have two X chomosomes, one from each parent. A male shares X chromosome genes with his maternal, but not his paternal, uncle.

predilections? No to all these questions. If you are proud, you can stay proud. If you prefer to be guilty, stay guilty. Nothing has changed. In explaining what I mean, I am less interested in this particular case than I am in using it to illustrate a more general point about genes and the bogey of genetic determinism.

There is an important distinction between a blueprint and a recipe.* A blueprint is a detailed, point-for-point specification of some end product like a house or a car. One diagnostic feature of a blueprint is that it is reversible. Give an engineer a car and he can reconstruct its blueprint. But offer to a chef a rival's *pièce de résistance* to taste and he will fail to reconstruct the recipe. There is a one-to-one mapping between components of a blueprint and components of the end product. This bit of the car corresponds to this bit of the blueprint. That bit of the car corresponds to that bit of the blueprint. There is no such one-to-one mapping in the case of a recipe. You can't isolate a particular blob of soufflé and seek one word of the recipe that 'determines' that blob. All the words of the recipe, taken together with all the ingredients, combine to form the whole soufflé.

Genes, in different aspects of their behaviour, are sometimes like blueprints and sometimes like recipes. It is important to keep the two aspects separate. Genes are digital, textual information, and they retain their hard, textual integrity as they change partners down the generations. Chromosomes – long strings of genes – are formally just like long computer tapes. When a portion of genetic tape is read in a cell, the first thing that happens to the information is that it is translated from one code to another: from the DNA code to a related code that dictates the exact shape of a protein molecule. So far, the gene behaves like a blueprint. There really is a one-to-one mapping between bits of gene and bits of protein, and it really is deterministic.

It is in the next step of the process – the development of a whole body and its psychological predispositions – that things start to get more complicated and recipe-like. There is seldom a simple one-to-one mapping between particular genes and 'bits' of body. Rather, there is a mapping between genes and rates at which processes happen during embryonic development. The eventual

*This distinction was also used in 'Darwin Triumphant' (p. 104).

effects on bodies and their behaviour are often multifarious and hard to unravel.

The recipe is a good metaphor but, as an even better one, think of the body as a blanket, suspended from the ceiling by 100,000 rubber bands, all tangled and twisted around one another. The shape of the blanket – the body – is determined by the tensions of all these rubber bands taken together. Some of the rubber bands represent genes, others environmental factors. A change in a particular gene corresponds to a lengthening or shortening of one particular rubber band. But any one rubber band is linked to the blanket only indirectly via countless connections amid the welter of other rubber bands. If you cut one rubber band, or tighten it, there will be a distributed shift in tensions, and the effect on the shape of the blanket will be complex and hard to predict.

In the same way, possession of a particular gene need not infallibly dictate that an individual will be homosexual. Far more probably the causal influence will be statistical. The effect of genes on bodies and behaviour is like the effect of cigarette smoke on lungs. If you smoke heavily, you increase the statistical odds that you'll get lung cancer. You won't infallibly give yourself lung cancer. Nor does refraining from smoking protect you infallibly from cancer. We live in a statistical world.

Imagine the following newspaper headline: 'Scientists discover that homosexuality is caused.' Obviously this is not news at all; it is trivial. Everything is caused. To say that homosexuality is caused by genes is more interesting, and it has the aesthetic merit of discomfiting politically-inspired bores, but it doesn't say more than my trivial headline does about the irrevocability of homosexuality.

Some genetic causes are hard to reverse. Others are easy. Some environmental causes are easy to reverse. Others are hard. Think how tenaciously we cling to the accent of childhood: an adult immigrant is labelled a foreigner for life. This is far more ineluctably deterministic than many genetic effects. It would be interesting to know the statistical likelihood that a child, subjected to a particular environmental influence such as religious indoctrination by nuns, will be able to escape the influence later on. It would similarly be interesting to know the statistical likelihood that a man possessing a particular gene in the Xq28 region of the X chromosome will turn out to be homosexual. The mere demonstration that there exists a gene 'for' homosexuality leaves the value of that likelihood almost totally open. Genes have no monopoly on determinism.

So, if you hate homosexuals or love them, if you want to lock them up or 'cure' them, your reasons had better have nothing to do with genes.

2.5

Son of Moore's Law[70]

Great achievers who have gone far sometimes amuse themselves by then going too far. Peter Medawar knew what he was doing when he wrote, in his review of James D. Watson's *The Double Helix*,

> It is simply not worth arguing with anyone so obtuse as not to realize that this complex of discoveries [molecular genetics] is the greatest achievement of science in the twentieth century.

Medawar, like the author of the book he was reviewing, could justify his arrogance in spades, but you don't have to be obtuse to dissent from his opinion. What about that earlier Anglo-American complex of discoveries known as the Neo-Darwinian Modern Synthesis? Physicists could make a good case for relativity or quantum mechanics, and cosmologists for the expanding universe. The 'greatest' anything is ultimately undecidable, but the molecular genetic revolution was undeniably one of the greatest achievements of science in the twentieth century – and that means of the human species, ever. Where shall we take it – or where will it take us – in the next fifty years? By mid-century, history may judge Medawar to have been closer to the truth than his contemporaries – or even he – allowed.

If asked to summarize molecular genetics in a word, I would choose 'digital'. Of course, Mendel's genetics was digital in being particulate with respect to the independent assortment of genes through pedigrees. But the interior of genes was unknown and they could still have been substances with continuously varying qualities, strengths and flavours, inextricably intertwined with their effects. Watson/Crick genetics is digital through and through, digital to its very backbone, the double helix itself. A genome's size can be measured in gigabases with exactly the same precision as a hard drive is sized up in gigabytes. Indeed, the two units are interconvertible by constant multiplication. Genetics today is pure information technology. This, precisely, is why an antifreeze gene can be copied from an Arctic fish and pasted into a tomato.*

The explosion sparked by Watson and Crick grew exponentially, as a good explosion should, during the half century since their famous joint publication. I think I mean that literally, and I'll support it by analogy with a better known explosion, this time from information technology as conventionally understood. Moore's Law states that computer power doubles every eighteen months. It is an empirical law without an agreed theoretical underpinning, though Nathan Myhrvold offers a wittily self-referential candidate: 'Nathan's Law' states that software grows faster than Moore's Law, and that is why we have Moore's Law. Whatever the underlying reason, or complex of reasons, Moore's Law has held true for nearly fifty years. Many analysts expect it to continue for as long again, with stunning effects upon human affairs – but that is not my concern in this essay.

Instead, is there something equivalent to Moore's Law for DNA information technology? The best measure would surely be an economic one, for money is a good composite index of man-hours and equipment costs. As the decades go by, what is the benchmark number of DNA kilobases that can be sequenced for a standard quantity of money? Does it increase exponentially, and if so what is its doubling time? Notice, by the way (it is another aspect of DNA science's being a branch of information technology) that it makes no difference which animal or plant provides the DNA. The sequencing techniques and the costs in any one decade are much the same. Indeed, unless you read the text message itself, it is impossible to tell whether DNA comes from a man, a mushroom or a microbe.

Having chosen my economic benchmark, I didn't know how to measure the costs in practice. Fortunately, I had the good sense to ask my colleague Jonathan Hodgkin, Professor of Genetics at Oxford University. I was delighted to discover that he had recently done the very thing while preparing a lecture for his old school, and he kindly sent me the following estimates of the cost, in pounds

*See 'Science, Genetics and Ethics: Memo for Tony Blair' (p. 32).

SON OF MOORE'S LAW

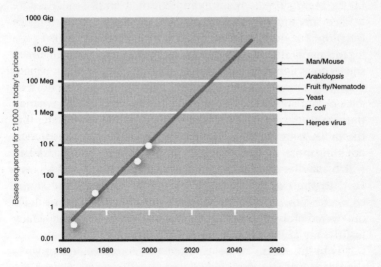

Linear regression fitted to four data points, then extrapolated to 2050

sterling, per base pair (that is, 'per letter' of the DNA code) sequenced. In 1965, it cost about £1000 per letter to sequence 5S ribosomal RNA from bacteria (not DNA, but RNA costs are similar). In 1975, to sequence DNA from the virus X174 cost about £10 per letter. Hodgkin didn't find a good example for 1985, but in 1995 it cost £1 per letter to sequence the DNA of *Caenorhabditis elegans*, the tiny nematode worm of which molecular biologists are so (rightly) enamoured that they call it 'the' nematode, or even 'the' worm.* By the time the Human Genome Project culminated around 2000, sequencing costs were about £0.1 per letter. To show

*The absurdity of this can be gauged from an image I have never forgotten, quoted in one of the first zoology books I ever owned, Ralph Buchsbaum's *Animals without Backbones* (University of Chicago Press). 'If all the matter in the universe except the nematodes were swept away, our world would still be dimly recognizable ... we should find its mountains, hills, vales, rivers, lakes, and oceans represented by a film of nematodes ... Trees would still stand in ghostly rows representing our streets and highways. The location of the various plants and animals would still be decipherable, and, had we sufficient knowledge, in many cases even their species could be determined by an examination of their erstwhile nematode parasites.' There are probably more than half a million species of nematodes, hugely outnumbering the species in all the vertebrate classes put together.

the positive trend of growth, I inverted these figures to 'bangs for the buck' – that is, quantity of DNA that can be sequenced for a fixed amount of money, and I chose £1000, correcting for inflation. I have plotted the resulting kilobases per £1000 on a logarithmic scale, which is convenient because exponential growth shows up as a straight line. (See graph on p. 129.)

I must emphasize, as Professor Hodgkin did to me, that the four data points are back-of-the-envelope calculations. Nevertheless, they do fall convincingly close to a straight line, suggesting that the increase in our DNA sequencing power is exponential. The doubling time (or cost-halving time) is twenty-seven months, which may be compared with the eighteen months of Moore's Law. To the extent that DNA sequencing work depends upon computer power (quite a large extent), the new law we have discovered probably owes a great deal to Moore's Law itself, which justifies my facetious label, 'Son of Moore's Law'.

It is by no means to be expected that technological progress should advance in this exponential way. I haven't plotted the figures out, but I'd be surprised if, say, speed of aircraft, fuel economy of cars, or height of skyscrapers were found to advance exponentially. Rather than double and double again in a fixed time, I suspect that they advance by something closer to arithmetic addition. Indeed, the late Christopher Evans, as long ago as 1979, when Moore's Law had scarcely begun, wrote:

> Today's car differs from those of the immediate postwar years on a number of counts ... But suppose for a moment that the automobile industry had developed at the same rate as computers and over the same period: how much cheaper and more efficient would the current models be? ... Today you would be able to buy a Rolls-Royce for £1.35*, it would do three million miles to the gallon, and it would deliver enough power to drive the *Queen Elizabeth II*. And if you were interested in miniaturization, you could place half a dozen of them on a pinhead.

Space exploration also seemed to me a likely candidate for modest additive increase like motor cars. Then I remembered a fascinating speculation mentioned by Arthur C. Clarke, whose credentials as

*Two US dollars.

a prophet are not to be ignored. Imagine a future spacecraft heading off for a distant star. Even travelling at the highest speed allowed by the current state of the art, it would still take many centuries to reach its distant destination. And before it had completed half its journey it would be overtaken by a faster vessel, the product of a later century's technology. So, it might be said, the original ship should never have bothered to set out. By the same argument, even the second spaceship should not bother to set out, because its crew is fated to wave to their great-grandchildren as they zoom by in a third. And so on. One way to resolve the paradox is to point out that the technology to develop later spaceships would not become available without the research and development that went into their slower predecessors. I would give the same answer to anybody who suggested that since the entire Human Genome Project could now be started from scratch and completed in a fraction of the years the actual project took, the original enterprise should have been postponed appropriately.

If our four data points are admittedly rough estimates, the extrapolation of the straight line out to the year 2050 is even more tentative. But by analogy with Moore's Law, and especially if Son of Moore's Law really does owe something to its parent, this straight line probably represents a defensible prognostication. Let's at least follow to see where it will take us. It suggests that in the year 2050 we shall be able to sequence a complete individual human genome for £100 at today's values (about $160). Instead of 'the' human genome project, every individual will be able to afford their own personal genome project. Population geneticists will have the ultimate data on human diversity. It will be possible to work out trees of cousinship linking any person in the world to any other person. It is a historian's wildest dream. They will use the geographic distribution of genes to reconstruct the great migrations and invasions of the centuries, track voyages of Viking longships, follow the American tribes by their genes down from Alaska to Tierra del Fuego and the Saxons across Britain, document the diaspora of the Jews, even identify the modern descendants of pillaging warlords like Genghis Khan.*

*DNA analysis is already making exciting contributions to historical research. See, for example, Bryan Sykes, *The Seven Daughters of Eve* (London, Bantam Press, 2001) and S. Wells, *The Journey of Man: A Genetic Odyssey* (London, Allen Lane, 2002).

Today, a chest X-ray will tell you whether you have lung cancer or tuberculosis. In 2050, for the price of a chest X-ray, you will be able to know the full text of every one of your genes. The doctor will hand you not the prescription recommended for an average person with your complaint but the prescription that precisely suits your genome. That is no doubt good, but your personal printout will also predict, with alarming precision, your natural end. Shall we want such knowledge? Even if we want it ourselves, shall we want our DNA printout to be read by insurance actuaries, paternity lawyers, governments? Even in a benign democracy, not everybody is happy with such a prospect. How some future Hitler might abuse this knowledge needs thinking about.

Weighty as such concerns may be, they are again not mine in this essay. I retreat to my ivory tower and more academic pre-occupations. If £100 becomes the price of sequencing a human genome, the same money will buy the genome of any other mammal; all are about the same size, in the gigabase order of magnitude, as is true of all vertebrates. Even if we assume that Son of Moore's Law will flatten off before 2050, as many people believe Moore's Law will, we can still safely predict that it will become economically feasible to sequence the genomes of hundreds of species per year. Having such a welter of information is one thing. What can we do with it? How shall we digest it, sift it, collate it, use it?

One relatively modest goal will be total and final knowledge of the phylogenetic tree. For there is, after all, one true tree of life, the unique pattern of evolutionary branching that actually happened. It exists. It is in principle knowable. We don't know it all yet. By 2050 we should – or if we do not, we shall have been defeated only at the terminal twigs, by the sheer number of species (a number that, as my colleague Robert May points out, is at present unknown to the nearest one or even two orders of magnitude).

My research assistant Yan Wong suggests that naturalists and ecologists in 2050 will carry a small field taxonomy kit, which will obviate the need to send specimens off to a museum expert for identification. A fine probe, hooked up to a portable computer, will be inserted into a tree, or a freshly trapped vole or grasshopper. Within minutes, the computer will chew over a few key segments of DNA, then spit out the species name and any other details that may be in its stored database.

Already, DNA taxonomy has turned up some sharp surprises. My traditional zoologist's mind protests almost unendurably at being asked to believe that hippos are more closely related to whales than they are to pigs. This is still controversial. It will be settled, one way or the other, along with countless other such disputes, by 2050. It will be settled because the Hippo Genome Project, the Pig Genome Project, and the Whale (if our Japanese friends haven't eaten them all by then) Genome Project will have been completed. Actually, it will not be necessary to sequence entire genomes to dissolve taxonomic uncertainty forever.

A spin-off benefit, which will perhaps have its greatest impact in the United States, is that full knowledge of the tree of life will make it even harder to doubt the fact of evolution. Fossils will become by comparison irrelevant to the argument, as hundreds of separate genes, in as many surviving species as we can bear to sequence, are found to corroborate each other's accounts of the one true tree of life.

It has been said often enough to become a platitude but I had better say it again: to know the genome of an animal is not the same as to understand that animal. Following Sydney Brenner (the single individual regarding whom, more than any other, I have heard people wonder at the absence so far of a Nobel Prize*), I shall think in terms of three steps, of increasing difficulty, in 'computing' an animal from its genome. Step 1 was hard but has now been completely solved. It is to compute the amino acid sequence of a protein from the nucleotide sequence of a gene. Step 2 is to compute the three-dimensional folding pattern of a protein from its one-dimensional sequence of amino acids. Physicists believe that in principle this can be done, but it is hard, and it may often be quicker to make the protein and see what happens. Step 3 is to compute the developing embryo from its genes and their interaction with their environment – which mostly consists of other genes. This is the hardest step, but the

*Stop press: Sydney Brenner's Nobel Prize was announced while this book was in proof.

science of embryology (especially of the workings of Hox and similar genes) is advancing at such a rate that by 2050 it will probably be solved. In other words, I conjecture that an embryologist of 2050 will feed the genome of an unknown animal into a computer, and the computer will simulate an embryology that will culminate in a full rendering of the adult animal. This will not be a particularly useful accomplishment in itself, since a real embryo will always be a cheaper computer than an electronic one. But it will be a way of signifying the completeness of our understanding. And particular implementations of the technology will be useful. For instance, detectives finding a bloodstain may be able to issue a computer image of the face of a suspect – or rather, since genes don't mature with age, a series of faces from babyhood to dotage!

I also think that by 2050 my dream of the Genetic Book of the Dead will become a reality. Darwinian reasoning shows that the genes of a species must constitute a kind of description of the ancestral environments through which those genes have survived. The gene pool of a species is the clay which is shaped by natural selection. As I put it in *Unweaving the Rainbow*:

> Like sandbluffs carved into fantastic shapes by the desert winds, like rocks shaped by ocean waves, camel DNA has been sculpted by survival in ancient deserts, and even more ancient seas, to yield modern camels. Camel DNA speaks – if only we could read the language – of the changing worlds of camel ancestors. If only we could read the language, the DNA of tuna and starfish would have 'sea' written into the text. The DNA of moles and earthworms would spell 'underground'.

I believe that by 2050 we shall be able to read the language. We shall feed the genome of an unknown animal into a computer which will reconstruct not only the form of the animal but the detailed world in which its ancestors (who were naturally selected to produce it) lived, including their predators or prey, parasites or hosts, nesting sites, and even hopes and fears.

What about more direct reconstructions of ancestors, Jurassic Park style? DNA in amber is, unfortunately, unlikely to be preserved intact, and no sons or even grandsons of Moore's Law are going to bring it back. But there probably are ways, many of them as yet scarcely dreamed of, by which we can use the copious data banks of surviving DNA that we shall have even before 2050. The Chimpanzee Genome Project is already under way and, thanks to Son of Moore's Law, should be completed in a fraction of the time taken by the human genome.

In a throwaway remark at the end of his own piece of millennial crystal-gazing,[71] Sydney Brenner made the following startling suggestion. When the chimpanzee genome is fully known, it should become possible, by a sophisticated and biologically intelligent comparison with the human genome (the two differ in only a tiny percentage of their DNA letters), to reconstruct the genome of the ancestor we share. This animal, the so-called 'missing link', lived between 5 million and 8 million years ago, in Africa. Once Brenner's leap is accepted, it is tempting to extend the reasoning all over the place, and I am not one to resist such temptation. The Missing Link Genome Project (MLGP) completed, the next step might be to line up the MLG with the human genome for a base-by-base comparison. Splitting the difference between the two (in the same kind of embryologically informed way as before) should yield a generalized approximation to *Australopithecus*, the genus of which Lucy has become the iconic representative. By the time the LGP (Lucy Genome Project) has been completed, embryology should have advanced to the point where the reconstructed genome could be inserted into a human egg and implanted in a woman, and a new Lucy born into the light of today. This will doubtless raise ethical worries.

Though concerned for the happiness of the individual australopithecine reconstructed (this is at least a coherent ethical issue, unlike fatuous worries about 'playing God'), I can see positive ethical benefits, as well as scientific ones, emerging from the experiment. At present we get away with our flagrant speciesism because the evolutionary intermediates between us and chimpanzees are all extinct. In my contribution to *The Great Ape Project* I pointed out that the accidental contingency of such extinction should be enough to destroy absolutist valuings of human life above all other life.[72] 'Pro life', for example, in debates on abortion or stem cell research, always means pro *human* life, for no sensibly articulated reason. The existence of a living, breathing Lucy in our

135

midst would change, forever, our complacent, human-centred view of morals and politics. Should Lucy pass for human? The absurdity of the question should be self-evident, as in those South African courts which tried to decide whether particular individuals should 'pass for white'. The reconstruction of a Lucy would be ethically vindicated by bringing such absurdity out into the open.

While the ethicists, moralists and theologians (I fear there still will be theologians in 2050) are busy agonizing over Project Lucy, biologists could, with relative impunity, be cutting their teeth on something even more ambitious: Project Dinosaur. And they might do it by, among other things, helping birds to cut teeth as they haven't done for 60 million years.

Modern birds are descended from dinosaurs (or at least from ancestors we would now happily call dinosaurs if only they had gone extinct as decent dinosaurs should). A sophisticated 'evo-devo' (evolution and development) interpretation of modern bird genomes and the genomes of other surviving archosaurian reptiles such as crocodiles might enable us, by 2050, to reconstruct the genome of a generalized dinosaur. It is encouraging already that a chicken beak can be experimentally induced to grow tooth buds (and snakes induced to grow legs), indicating that ancient genetic skills still linger. If the Dinosaur Genome Project is successful, we could perhaps implant the genome in an ostrich egg to hatch a living, breathing, terrible lizard. Jurassic Park notwithstanding, my only anxiety is that I am unlikely to live long enough to see it. Or to extend my short arm to a new Lucy's long one and shake her tearfully by the hand.

3

THE INFECTED MIND

I have long been academically attracted, and humanly repelled, by the idea that self-replicating information leaps infectiously from mind to mind like (what we now know as) computer viruses. Whether or not we use the name 'meme' for these mind viruses, the theory needs to be taken seriously. If rejected, it must be rejected for good reasons. One of those who have taken it very seriously is Susan Blackmore, in her admirable book, *The Meme Machine*. The first essay in this section, **Chinese Junk and Chinese Whispers** (3.1), is a shortened version of my Foreword to her book. I used the opportunity to think afresh about memes, and I concluded by rebutting the suggestion that I have gone cold on memes since introducing them in 1976. As with other Forewords to books, those parts which were concerned specifically with the book itself have been cut, not because I no longer stand by them (I do), but because they are too particular for a collection such as this.

From 1976 onwards, I always thought religions provided the prime examples of memes and meme complexes (or 'memeplexes'). In **Viruses of the Mind** (3.2) I developed this theme of religions as mind parasites, and also the analogy with computer viruses. It first appeared in an edited book of responses to the thinking of Daniel Dennett, a philosopher of science whom scientists like because he bothers to read science. My choice of topic acknowledged Dennett's fertile development of the meme concept in *Consciousness Explained* and *Darwin's Dangerous Idea*.[73]

To describe religions as mind viruses is sometimes interpreted as contemptuous or even hostile. It is both. I am often asked why I am so hostile to 'organized religion'. My first response is that I am not exactly friendly towards disorganized religion either. As a lover of truth, I am suspicious of strongly held beliefs that are unsupported by evidence:

fairies, unicorns, werewolves, any of the infinite set of conceivable and unfalsifiable beliefs epitomized by Bertrand Russell's hypothetical china teapot orbiting the Sun (see 'The Great Convergence', pp. 177–78). The reason organized religion merits outright hostility is that, unlike belief in Russell's teapot, religion is powerful, influential, tax-exempt and systematically passed on to children too young to defend themselves.* Children are not compelled to spend their formative years memorizing loony books about teapots. Government-subsidized schools don't exclude children whose parents prefer the wrong shape of teapot. Teapot-believers don't stone teapot-unbelievers, teapot-apostates, teapot-heretics and teapot-blasphemers to death. Mothers don't warn their sons off marrying teapot-shiksas whose parents believe in three teapots rather than one. People who put the milk in first don't kneecap those who put the tea in first.

The rest of this section is all about religion, not specifically the viral analogy, although that is always in my mind when I consider religion.† **The Great Convergence** (3.3) discusses, and rejects, a fashionable claim that science and religion, having drifted apart, are now coming together again. **Dolly and the Cloth Heads** (3.4) criticizes the tendency for decent, liberal societies, and especially our public media, to grant religious spokesmen a privileged platform, and an exaggerated respect which goes beyond that due them as individuals. It is a general complaint, but the particular stimulus for this article was Dolly the charismatic sheep. Of course theologians are as entitled as anybody else to hold opinions on such matters. What I objected to was only the automatic, unquestioned assumption that opinions should be given an inside track to our attention simply because they come from religion.

The attack on automatic respect continues in the next essay, **Time**

---

*See page 151 and also Nicholas Humphrey's brilliant Amnesty Lecture, 'What shall we tell the children?', originally published in W. Williams (ed.), *The Values of Science: The Oxford Amnesty Lectures 1997* (Boulder, Westview Press, 1999) and now reprinted in Humphrey's collection of essays, *The Mind Made Flesh* (Oxford, Oxford University Press, 2002).

†Which is not to imply that the viral theory, on its own, suffices to explain the phenomenon of religion. Two thoughtful books that have taken a biological, or psychological, approach to the question are Robert Hinde, *Why Gods Persist* (London, Routledge, 1999) and Pascal Boyer, *Religion Explained* (London, Heinemann, 2001).

to Stand Up (3.5). I wrote it in the immediate aftermath of the religious atrocity committed in New York on 11 September 2001, and it has a more savage tone than I customarily adopt. Were I to rewrite it now, I should probably tone it down, but that was an extraordinary time when people spoke with extraordinary passion, and I admit that I was no exception.

Chinese Junk and Chinese Whispers[74]

From the Foreword to *The Meme Machine*
by Susan Blackmore

As an undergraduate I was chatting to a friend in the college lunch queue. He regarded me with increasingly quizzical amusement, then asked: 'Have you just been with Peter Brunet?' I had indeed, though I couldn't guess how he knew. Peter Brunet was our much loved tutor, and I had come hot foot from a tutorial hour with him. 'I thought so,' my friend laughed. 'You are talking just like him; your voice sounds exactly like his.' I had, if only briefly, 'inherited' intonations and manners of speech from an admired, and now greatly missed, teacher.

Years later, when I became a tutor myself, I taught a young woman who affected an unusual habit. When asked a question which required deep thought, she would screw her eyes tight shut, jerk her head down to her chest and then freeze for up to half a minute before looking up, opening her eyes, and answering the question with fluency and intelligence. I was amused by this, and did an imitation of it to divert my colleagues after dinner. Among them was a distinguished Oxford philosopher. As soon as he saw my imitation, he immediately said: 'That's Wittgenstein! Is her surname —— by any chance?' Taken aback, I said that it was. 'I thought so,' said my colleague. 'Both her parents are devoted followers of Wittgenstein.' The gesture had passed from the great philosopher, via one or both of her parents, to my pupil. I suppose that, although my further imitation was done in jest, I must count myself a fourth generation transmitter of the gesture. And who knows where Wittgenstein got it?

Imitation is how a child learns its particular language rather than some other language. It is why people speak more like their own parents than like other people's parents. It is why regional accents, and on a longer timescale separate languages, exist. It is why religions persist along family lines rather than being chosen afresh in every generation. There is at least a superficial analogy to the longitudinal transmission of genes down generations, and to the horizontal transmission of genes in viruses. Without prejudging the issue of whether the analogy is a fruitful one, if we want even to talk about it we had better have a name for the entity that might play the role of gene in the transmission of words, ideas, faiths, mannerisms and fashions. Since 1976, when the word was coined, increasing numbers of people have adopted the name 'meme' for the postulated gene analogue.

The compilers of the Oxford English Dictionaries operate a sensible criterion for deciding whether a new word shall be canonized by inclusion. The aspirant word must be commonly used without needing to be defined and without its coining being attributed. To ask the metamemetic question, how widespread is 'meme', a far from ideal, but nevertheless convenient method of sampling the meme pool, is provided by the World Wide Web. I did a quick search of the web on the day of writing this, which happened to be 29 August 1998. 'Meme' is mentioned about half a million times, but that's a ridiculously high figure, obviously confounded by various acronyms and the French *même*. The adjectival form 'memetic' is genuinely exclusive, and it clocked up 5042 mentions. To put this number into perspective, I compared a few other recently coined words or fashionable expressions. Spin doctor (or spin-doctor) gets 1412 mentions, dumbing down 3905, docudrama (or docu-drama) 2848, sociobiology 6679, catastrophe theory 1472, edge of chaos 2673, wannabee 2650, zippergate 1752, studmuffin 776, post-structural (or poststructural) 577, extended phenotype 515, exaptation 307. Of the 5042 mentions of memetic, more than 90 per cent make no mention of the origin of the word, which suggests that it does indeed meet the OED's criterion. And the Oxford Dictionary now does contain the following definition: **meme**: 'a self-replicating element of culture, passed on by imitation.'

Further searching of the internet reveals a newsgroup talking-shop, 'alt.memetics', which has received about 12,000 postings during the past year. There are on-line articles on, among many other things, 'The New Meme', 'Meme, Counter-meme', 'Memetics: a Systems Metabiology', 'Memes, and Grinning Idiot Press', 'Memes, Metamemes and Politics', 'Cryonics, religions and memes', 'Selfish Memes and the evolution of cooperation', and 'Running down the Meme'. There are separate web pages on 'Memetics', 'Memes', 'The C Memetic Nexus', 'Meme theorists on the web', 'Meme of the week', 'Meme Central', 'Arkuat's Meme Workshop', 'Some pointers and a short introduction to memetics', 'Memetics Index' and 'Meme Gardening Page'. There is even a new religion (tongue in cheek, I *think*), called the 'Church of Virus', complete with its own list of Sins and Virtues, and its own patron saint (Saint Charles Darwin, canonized as 'perhaps the most influential memetic engineer of the modern era') and I was alarmed to discover a passing reference to 'Saint Dawkins'.

Memes travel longitudinally down generations, but they travel horizontally too, like viruses in an epidemic. Indeed, it is largely horizontal epidemiology that we are studying when we measure the spread of a word like memetic, docudrama or studmuffin over the internet. Crazes among schoolchildren provide particularly tidy examples. When I was about nine, my father taught me to fold a square of paper to make an origami Chinese junk. It was a remarkable feat of artificial embryology, passing through a distinctive series of intermediate stages: catamaran with two hulls, cupboard with doors, picture in a frame, and finally the junk itself, fully seaworthy or at least bathworthy, complete with deep hold, and two flat decks each surmounted by a large, square-rigged sail. The point of the story is that I went back to school and infected my friends with the skill, and it then spread around the school with the speed of the measles and pretty much the same epidemiological time-course. I don't know whether the epidemic subsequently jumped to other schools (a boarding school is a somewhat isolated backwater of the meme pool). But I do know that my father himself originally picked up the Chinese Junk meme during an almost identical epidemic at the same school 25 years earlier. The earlier virus was launched by the school matron. Long after the old matron's departure, I had reintroduced her meme to a new cohort of small boys.

Before leaving the Chinese junk, let me use it to make one more point. A favourite objection to the meme/gene analogy is that memes, if they exist at all, are transmitted with too low fidelity to perform a gene-like role in any realistically Darwinian selection process. The difference between high fidelity genes and low fidelity memes is assumed to follow from the fact that genes, but not memes, are digital. I am sure that the details of Wittgenstein's mannerism were far from faithfully reproduced when I imitated my pupil's imitation of her parents' imitation of Wittgenstein. The form and timing of the tic undoubtedly mutated over the generations, as in the childhood game of Chinese Whispers (Americans call it Telephone).

Suppose we assemble a line of children. A picture of, say, a Chinese junk is shown to the first child, who is asked to draw it. The drawing, but not the original picture, is then shown to the second child, who is asked to make her own drawing of it. The second child's drawing is shown to the third child, who draws it again, and so the series proceeds until the twentieth child, whose drawing is revealed to everyone and compared with the first. Without even doing the experiment, we know what the result will be. The twentieth drawing will be so unlike the first as to be unrecognizable. Presumably, if we lay the drawings out in order, we shall notice some resemblance between each one and its immediate predecessor and successor, but the mutation rate will be so high as to destroy all semblance after a few generations. A trend will be visible as we walk from one end of the series of drawings to the other, and the direction of the trend will be degeneration. Evolutionary geneticists have long understood that natural selection cannot work unless the mutation rate is low. Indeed, the initial problem of overcoming the fidelity barrier has been described as the Catch-22 of the Origin of Life. Darwinism depends upon high fidelity gene replication. How then can the meme, with its apparently dismal lack of fidelity, serve as quasi-gene in any quasi-Darwinian process?

It isn't always as dismal as you think, and high fidelity is not necessarily synonymous with digital. Suppose we set up our

Chinese Whispers game again, but this time with a crucial difference. Instead of asking the first child to copy a drawing of a junk, we teach her, by demonstration, to make an origami model of a junk. When she has mastered the skill and made her own junk, the first child is asked to turn round to the second child and teach him how to make one. So the skill passes down the line to the twentieth child. What will be the result of this experiment? What will the twentieth child produce, and what shall we observe if we lay the 20 efforts out in order along the ground? I haven't done it, but I will make the following confident prediction, assuming that we run the experiment many times on different groups of 20 children. In several of the experiments, a child somewhere along the line will forget some crucial step in the skill taught him by the previous child, and the line of phenotypes will suffer an abrupt macromutation which will presumably then be copied to the end of the line, or until another discrete mistake is made. The end result of such mutated lines will not bear any resemblance to a Chinese junk at all. But in a good number of experiments the skill will correctly pass all along the line, and the twentieth junk will be no worse and no better, on average, than the first junk. If we then lay the 20 junks out in order, some will be more perfect than others, but imperfections will not be copied on down the line. If the fifth child is hamfisted and makes a clumsily asymmetrical or floppy junk, his quantitative errors will be corrected if the sixth child happens to be more dexterous. The 20 junks will not exhibit a progressive deterioration in the way that the 20 drawings of our first experiment undoubtedly would.

Why? What is the crucial difference between the two kinds of experiment? It is this. Inheritance in the drawing experiment is Lamarckian (Susan Blackmore calls it 'copying the product'). In the origami experiment it is Weismannian (Blackmore's 'copying the instructions'). In the drawing experiment, the phenotype in every generation is also the genotype – it is what is passed on to the next generation. In the origami experiment, what passes to the next generation is not the paper phenotype but a set of instructions for making it. Imperfections in the *execution* of the instructions result in imperfect junks (phenotypes) but they are not passed on to future generations: they are non-memetic. Here are the first five instructions in the Weismannian meme-line of instructions for making a Chinese junk:

1. Take a square sheet of paper and fold all four corners exactly into the middle.
2. Take the reduced square so formed, and fold one side into the middle.
3. Fold the opposite side into the middle, symmetrically.
4. In the same way, take the rectangle so formed, and fold its two ends into the middle.
5. Take the small square so formed, and fold it backwards, exactly along the straight line where your last two folds met.

... And so on, through 20 or 30 instructions of this kind. These instructions, though I would not wish to call them digital, are potentially of very high fidelity, just as if they were digital. This is because they all make reference to idealized tasks like 'fold the four corners exactly into the middle'. If the paper is not exactly square, or if a child folds ineptly so that, say, the first corner overshoots the middle and the fourth corner undershoots it, the junk that results will be inelegant. But the next child in the line will not copy the error, for she will assume that her instructor *intended* to fold all four corners into the exact centre of a perfect square. The instructions are self-normalizing. The code is error-correcting.

The instructions are more effectively passed on if verbally reinforced, but they can be transmitted by demonstration alone. A Japanese child could teach an English one, though neither has a word of the other's language. In the same way, a Japanese master carpenter could convey his skills to an equally monoglot English apprentice. The apprentice would not copy obvious mistakes. If the master hit his thumb with a hammer, the apprentice would correctly guess, even without understanding the Japanese for '** **** **!', that he meant to hit the nail. He would not make a Lamarckian copy of the precise details of every hammer blow, but copy instead the inferred instruction: drive the nail in with as many blows of your hammer as it takes your arm to achieve the same idealized end result as the master has achieved with his – a nail-head flush with the wood.

I believe that these considerations greatly reduce, and probably remove altogether, the objection that memes are copied with insufficient fidelity to be compared with genes. For me, the quasi-genetic inheritance of language, and of religious and traditional customs, teaches the same lesson. Another objection is that we don't know what memes are made of or where they reside. Memes have not yet found their Watson and Crick; they even lack their Mendel. Whereas genes are to be found in precise locations on chromosomes, memes presumably exist in brains, and we have even less chance of seeing one than of seeing a gene (though the neurobiologist Juan Delius has pictured his conjecture of what a meme might look like[75]). As with genes, we track memes through populations by their phenotypes. The 'phenotype' of the Chinese junk meme is made of paper. With the exception of 'extended phenotypes' such as beaver dams and caddis larva houses, the phenotypes of genes are normally parts of living bodies. Meme phenotypes seldom are.

But it can happen. To return to my school again, a Martian geneticist, visiting the school during the morning cold bath ritual, would have unhesitatingly diagnosed an 'obvious' genetic polymorphism. About 50 per cent of the boys were circumcised and 50 per cent were not. The boys, incidentally, were highly conscious of the polymorphism and we classified ourselves into Roundheads versus Cavaliers (I have recently read of another school in which the boys even organized themselves into two football teams along the same lines). It is, of course, not a genetic but a memetic polymorphism. But the Martian's mistake is completely understandable; the morphological discontinuity is of exactly the kind that one normally expects to find produced by genes.

In England at the time, infant circumcision was a medical whim, and the roundhead/cavalier polymorphism at my school probably owed less to longitudinal transmission than to differing fashions in the various hospitals where we happened to have been born – horizontal memetic transmission yet again. But through most of history circumcision has been longitudinally transmitted as a badge of religion (of *parents'* religion I hasten to point out, for the unfortunate child is normally too young to *know* his own religious mind). Where circumcision is religiously or traditionally based (the barbaric custom of female 'circumcision' always is), the transmission will follow a longitudinal pattern of heredity, very similar to the pattern for true genetic transmission, and often persisting for many generations. Our Martian geneticist would have to work quite hard to discover that no genes are involved in the genesis of the roundhead phenotype.

The Martian geneticist's eyes would also pop out on stalks (assuming they weren't on stalks to begin with) at the contemplation of certain styles of clothing and hairdressing, and their inheritance patterns. The black skullcapped phenotype shows a marked tendency towards longitudinal transmission from father to son (or it may be from maternal grandfather to grandson), and there is clear linkage to the rarer pigtail-plaited sideburn phenotype. Behavioural phenotypes such as genuflecting in front of crosses, and facing east to kneel five times per day, are inherited longitudinally too, and are in strong linkage disequilibrium with the previously mentioned phenotypes, as is the red-dot-on-forehead phenotype, and the saffron robes/shaven head linkage group.

Genes are accurately copied and transmitted from body to body, but some are transmitted at greater frequency than others – by definition they are more successful. This is natural selection, and it is the explanation for most of what is interesting and remarkable about life. But is there a similar meme-based natural selection? Perhaps we can use the internet again to investigate natural selection among memes? As it happens, around the time the word meme was coined (actually a little later), a rival synonym, 'culturgen', was proposed.[76] Today, culturgen is mentioned 20 times on the World Wide Web, compared with meme's 5042. Moreover, of those 20, 17 also mention the source of the word, falling foul of the Oxford Dictionary's criterion. Perhaps it is not too fanciful to imagine a Darwinian struggle between the two memes (or culturgens), and it is not totally silly to ask why one of them was so much more successful. Perhaps it is because meme is a monosyllable similar to gene, which therefore lends itself to quasi-genetic sub-coinings: meme pool (352), memotype (58), memeticist (163), memeoid (or memoid) (28), retromeme (14), population memetics (41), meme complex (494), memetic engineering (302) and metameme (71) are all listed in a 'Memetic Lexicon' on the World Wide Web (the numbers in brackets count the mentions of each word on the World Wide Web on my sampling day). Culturgen-based equivalents would be less snappy. Or the success of meme against culturgen may have been initially just a non-Darwinian matter of chance – memetic drift (85) – followed by a self-reinforcing positive feedback effect ('unto every one that hath shall be given, and he shall have abundance: but from him that hath not shall be taken away even that which he hath' (Matthew 25:29).

I have mentioned two favourite objections to the meme idea: memes have insufficient copying fidelity, and nobody really knows what a meme physically is. A third is the vexed question of how large a unit deserves the name meme. Is the whole Roman Catholic Church one meme, or should we use the word for one constituent unit, such as the idea of incense or the transubstantiation? Or for something in between? The answer is to be found in the concept of the meme-complex or 'memeplex'.

Memes, like genes, are selected against the background of other memes in the meme pool. The result is that gangs of mutually compatible memes – coadapted meme complexes or memeplexes – are found cohabiting in individual brains. This is not because selection has chosen them as a group, but because each separate member of the group tends to be favoured when its environment happens to be dominated by the others. An exactly similar point can be made about genetic selection. Every gene in a gene pool constitutes part of the environmental background against which the other genes are naturally selected, so it's no wonder natural selection favours genes that 'cooperate' in building those highly integrated and unified machines called organisms. By analogy with coadapted gene complexes, memes, selected against the background of each other, 'cooperate' in mutually supportive memeplexes – supportive within the memeplex but hostile to rival memeplexes. Religions may be the most convincing examples of memeplexes, but they are by no means the only ones.

I am occasionally accused of having backtracked on memes; of having lost heart, pulled in my horns, had second thoughts. The truth is that my first thoughts were more modest than some memeticists might have wished. For me, the original mission of the meme was negative. The word was introduced at the end of a book which otherwise must have seemed entirely devoted to extolling the selfish gene as the be-all and end-all of evolution, the fundamental unit of selection, the entity in the hierarchy of life which all adaptations could be said to benefit. There was a risk that my readers would misunderstand the message as being *necessarily* about genes in the sense of DNA molecules. On the contrary, DNA was incidental. The real unit of natural selection was any kind of *replicator*, any unit of which copies are made, with occasional errors, and with some influence or power over their own probability of replication. The genetic natural selection identified by neo-Darwinism as the driving force of evolution on this planet was only a special case of a more general process that I came to dub 'Universal Darwinism'. Perhaps we'd have to go to other planets in order to discover any other examples. But maybe we didn't have to go that far. Could it be that a new kind of Darwinian replicator was even now staring us in the face? This was where the meme came in.

I would have been content, then, if the meme had done its work of simply persuading my readers that the gene was only a special case: that its role in the play of Universal Darwinism could be filled by any entity in the universe answering to the definition of Replicator. The original didactic purpose of the meme was the negative one of cutting the selfish gene down to size. I became a little alarmed at the number of my readers who took the meme more positively as a theory of human culture in its own right – either to criticize it (unfairly, given my original modest intention) or to carry it far beyond the limits of what I then thought justified. This was why I may have seemed to backtrack.

But I was always open to the possibility that the meme might one day be developed into a proper hypothesis of the human mind, and I did not know how ambitious such a thesis might turn out to be. I am delighted that others are now undertaking it.*

---

*In addition to Susan Blackmore's *The Meme Machine*, other books that make heavy use of the meme idea are R. Brodie, *Virus of the Mind: the New Science of the Meme* (Seattle, Integral Press, 1996) (not to be confused with my essay (see opposite page), which was published three years earlier); A. Lynch, *Thought Contagion: How Belief Spreads Through Society* (New York, Basic Books, 1998); J. M. Balkin, *Cultural Software* (New Haven, Yale University Press, 1998); H. Bloom, *The Lucifer Principle* (Sydney, Allen & Unwin, 1995); Robert Aunger, *The Electric Meme* (New York, Simon & Schuster, 2002); Kevin Laland and Gillian Brown, *Sense and Nonsense* (Oxford, Oxford University Press, 2002); and Stephen Shennan, *Genes, Memes and Human History* (London, Thames and Hudson, 2002). A turning point in the fortunes of the meme was its adoption and development by Daniel Dennett as a cornerstone of his theory of the evolution of the mind, especially in his two books *Consciousness Explained* (Boston, Little Brown, 1991) and *Darwin's Dangerous Idea* (New York, Simon & Schuster, 1995).

3.2

Viruses of the Mind[77]

The haven all memes depend on reaching is the human mind, but a human mind is itself an artifact created when memes restructure a human brain in order to make it a better habitat for memes. The avenues for entry and departure are modified to suit local conditions, and strengthened by various artificial devices that enhance fidelity and prolixity of replication: native Chinese minds differ dramatically from native French minds, and literate minds differ from illiterate minds. What memes provide in return to the organisms in which they reside is an incalculable store of advantages – with some Trojan horses thrown in for good measure …

Daniel Dennett[78]

Duplication-Fodder

A beautiful child close to me, six and the apple of her father's eye, believes that Thomas the Tank Engine really exists. She believes in Father Christmas, and when she grows up her ambition is to be a tooth fairy. She and her schoolfriends believe the solemn word of respected adults that tooth fairies and Father Christmas really exist. This little girl is of an age to believe whatever you tell her. If you tell her about witches changing princes into frogs, she will believe you. If you tell her that bad children roast forever in hell, she will have nightmares. I have just discovered that without her father's consent this sweet, trusting, gullible six-year-old is being sent, for weekly instruction, to a Roman Catholic nun. What chance has she?

A human child is shaped by evolution to soak up the culture of her people. Most obviously, she learns the essentials of their language in a matter of months. A large dictionary of words to speak, an encyclopaedia of information to speak about, complicated syntactic and semantic rules to order the speaking, all are transferred from older brains into hers well before she reaches half her adult size. When you are preprogrammed to absorb useful information at a high rate, it is hard to shut out pernicious or damaging information at the same time. With so many mind-bytes to be downloaded, so many mental codons to be duplicated, it is no wonder that child brains are gullible, open to almost any suggestion, vulnerable to subversion, easy prey to Moonies, Scientologists and nuns. Like immune-deficient patients, children are wide open to mental infections that adults might brush off without effort.

DNA, too, includes parasitic code. Cellular machinery is extremely good at copying DNA. Where DNA is concerned, it seems to have an eagerness to copy, like a child's eagerness to imitate the language of its parents. Concomitantly, DNA seems eager to be copied. The cell nucleus is a paradise for DNA, humming with sophisticated, fast and accurate duplicating machinery.

Cellular machinery is so friendly towards DNA-duplication that it is small wonder cells play host to DNA parasites – viruses, viroids, plasmids and a riff-raff of other genetic fellow travellers. Parasitic DNA even gets itself spliced seamlessly into the chromosomes themselves. 'Jumping genes' and stretches of 'Selfish DNA' cut or copy themselves out of chromosomes and paste themselves in elsewhere. Deadly oncogenes are almost impossible to distinguish from the legitimate genes between which they are spliced. In evolutionary time, there is probably a continual traffic from 'straight' genes to 'outlaw', and back again. DNA is just DNA. The only thing that distinguishes viral DNA from host DNA is its expected method of passing into future generations. 'Legitimate' host DNA is just DNA that aspires to pass into the next generation via the orthodox route of sperm or egg. 'Outlaw' or parasitic DNA is just DNA that looks to a quicker, less cooperative route to the future, via a sneezed droplet or a smear of blood, rather than via a sperm or egg.

For data on a floppy disk, a computer is a humming paradise just as cell nuclei hum with eagerness to duplicate DNA.

Computers and their associated disk and tape readers are designed with high fidelity in mind. As with DNA molecules, magnetized bytes don't literally 'want' to be faithfully copied. Nevertheless, you can write a computer program that takes steps to duplicate itself. Not just duplicate itself within one computer but spread itself to other computers. Computers are so good at copying bytes, and so good at faithfully obeying the instructions contained in those bytes, that they are sitting ducks to self-replicating programs: wide open to subversion by software parasites. Any cynic familiar with the theory of selfish genes and memes would have known that modern personal computers, with their promiscuous traffic of floppy disks and email links, were just asking for trouble. The only surprising thing about the current epidemic of computer viruses is that it has been so long in coming.

Computer Viruses: a Model for an Informational Epidemiology

Computer viruses are pieces of code that graft themselves into existing, legitimate programs and subvert the normal actions of those programs. They may travel on exchanged floppy disks, or over networks. They are technically distinguished from 'worms' which are whole programs in their own right, usually travelling over networks. Rather different are 'Trojan horses', a third category of destructive programs, which are not in themselves self-replicating but rely on humans to replicate them because of their pornographic or otherwise appealing content. Both viruses and worms are programs that actually say, in computer language, 'Duplicate Me'. Both may do other things that make their presence felt and perhaps satisfy the hole-in-corner vanity of their authors. These side effects may be 'humorous' (like the virus that makes the Macintosh's built-in loudspeaker enunciate the words 'Don't panic', with predictably opposite effect); malicious (like the viruses that erase the hard disk after a sniggering screen-announcement of the impending disaster); political (the Spanish Telecom and Beijing viruses protest about telephone costs and massacred students respectively); or simply inadvertent (the programmer is incompetent to handle the low-level system calls required to write an effective virus or worm). The famous Internet

Worm, which paralysed much of the computing power of the United States on 2 November 1988, was not intended (very) maliciously but got out of control and, within 24 hours, had clogged around 6000 computer memories with exponentially multiplying copies of itself.

> Memes now spread around the world at the speed of light, and replicate at rates that make even fruit flies and yeast cells look glacial in comparison. They leap promiscuously from vehicle to vehicle, and from medium to medium, and are proving to be virtually unquarantinable. [Dennett again]

Computer viruses aren't limited to electronic media such as disks and data lines. On its way from one computer to another, a virus may pass through printing ink, light rays in a human lens, optic nerve impulses and finger muscle contractions. A computer fanciers' magazine that printed the text of a virus program for the interest of its readers has been widely condemned. Indeed, such is the appeal of the virus idea to a certain kind of puerile mentality (the masculine gender is used advisedly), that publication of any kind of 'How to' information on designing virus programs is rightly seen as an irresponsible act.

I am not going to publish any virus code. But there are certain tricks of effective virus design that are sufficiently well known, even obvious, that it will do no harm to mention them, as I need to do in order to develop my theme. They all stem from the virus's need to evade detection while it is spreading.

A virus that clones itself too prolifically within one computer will soon be detected because the symptoms of clogging will become too obvious to ignore. For this reason many virus programs check, before infecting a system, to make sure that they are not already on that system. Incidentally, this opens the way for a defence against viruses that is analogous to immunization. In the days before a specific anti-virus program was available, I myself responded to an early infection of my own hard disk by means of a crude 'vaccination'. Instead of deleting the virus that I had detected, I simply disabled its coded instructions, leaving the 'shell' of the virus with its characteristic external 'signature' intact. In theory, subsequent members of the same virus species that arrived in my system should have recognized the signature of their own kind and refrained from trying to double-infect. I don't know whether this immunization really worked, but in those days it probably was worthwhile 'gutting' a virus and leaving a shell like this, rather than simply removing it lock, stock and barrel. Nowadays it is better to hand the problem over to one of the professionally written anti-virus programs.

A virus that is too virulent will be rapidly detected and scotched. A virus that instantly and catastrophically sabotages every computer in which it finds itself will not find itself in many computers. It may have a most amusing effect on one computer – erase an entire doctoral thesis or something equally side-splitting – but it won't spread as an epidemic. Some viruses, therefore, are designed to have an effect that is small enough to be difficult to detect, but which may nevertheless be extremely damaging. There is one type which, instead of erasing disk sectors wholesale, attacks only spreadsheets, making a few random changes in the (usually financial) quantities entered in the rows and columns. Other viruses evade detection by being triggered probabilistically, for example erasing only one in 16 of the hard disks infected. Yet other viruses employ the time-bomb principle. Most modern computers are 'aware' of the date, and viruses have been triggered to manifest themselves all around the world, on a particular date such as Friday 13th or April Fool's Day. From the parasitic point of view, it doesn't matter how catastrophic the eventual attack is, provided the virus has had plenty of opportunity to spread first (a disturbing analogy to the Medawar/Williams theory of ageing; we are the victims of lethal and sub-lethal genes that mature only after we have had plenty of time to reproduce). In defence, some large companies go so far as to set aside one 'miner's canary' among their fleet of computers, and advance its internal calendar a week so that any time-bomb viruses will reveal themselves prematurely before the big day.

Again predictably, the epidemic of computer viruses has triggered an arms race. Antiviral software is doing a roaring trade. These antidote programs – 'Interferon', 'Vaccine', 'Gatekeeper' and others – employ a diverse armoury of tricks. Some are written with specific, known and named, viruses in mind. Others intercept any attempt to meddle with sensitive system areas of memory and warn the user.

The virus principle could in theory be used for non-malicious, even beneficial purposes. Harold Thimbleby[79] coins the phrase 'Liveware' for his already-implemented use of the infection principle for keeping multiple copies of databases up to date. Every time a disk containing the database is plugged into a computer, it looks to see whether there is already another copy present on the local hard disk. If there is, each copy is updated in the light of the other. So, with a bit of luck, it doesn't matter which member of a circle of colleagues enters, say, a new bibliographic citation on his personal disk. His newly entered information will readily infect the disks of his colleagues (because the colleagues promiscuously insert their disks into one another's computers) and will spread like an epidemic around the circle. Thimbleby's liveware is not entirely virus-like: it could not spread to just anybody's computer and do damage. It spreads data only to already-existing copies of its own database; and you will not be infected by liveware unless you positively opt for infection.

Incidentally, Thimbleby, who is much concerned with the virus menace, points out that you can gain some protection by using computer systems that other people don't use. The usual justification for purchasing today's numerically dominant personal computer is simply and solely that it *is* numerically dominant. Almost every knowledgeable person agrees that, in terms of quality and especially user-friendliness, the rival, minority system is superior. Nevertheless, ubiquity is held to be a good in itself, sufficient to outweigh sheer quality. Buy the same (albeit inferior) computer as your colleagues, the argument goes, and you'll be able to benefit from shared software, and from a generally larger circulation of available software. The irony is that, with the advent of the virus plague, 'benefit' is not all that you are likely to get. Not only should we all be very hesitant before we accept a disk from a colleague, we should also be aware that, if we join a large community of users of a particular make of computer, we are also joining a larger community of viruses – even, it turns out, *disproportionately* larger.

Returning to possible uses of viruses for positive purposes, there are proposals to exploit the 'poacher turned gamekeeper' principle, and 'set a thief to catch a thief'. A simple way would be to take any of the existing antiviral programs and load it, as a 'warhead', into a harmless self-replicating virus. From a 'public health' point of view a spreading epidemic of antiviral software could be especially beneficial because the computers most vulnerable to malicious viruses – those whose owners are promiscuous in the exchange of pirated programs – will also be most vulnerable to infection by the healing antivirus. A more penetrating antivirus might – as in the immune system – 'learn' or 'evolve' an improved capacity to attack whatever viruses it encountered.

I can imagine other uses of the computer virus principle which, if not exactly altruistic, are at least constructive enough to escape the charge of pure vandalism. A computer company might wish to do market research on the habits of its customers, with a view to improving the design of future products. Do users like to choose files by pictorial icon, or do they opt to display them by textual name only? How deeply do people nest folders (directories) within one another? Do people settle down for a long session with only one program, say a word processor, or are they constantly switching back and forth, say between writing and drawing programs? Do people succeed in moving the mouse pointer straight to the target, or do they meander around in time-wasting hunting movements that could be rectified by a change in design?

The company could send out a questionnaire asking all these questions, but the customers that replied would be a biased sample and, in any case, their own assessment of their computer-using behaviour might be inaccurate. A better solution would be a market research computer program. Customers would be asked to load this program into their system where it would unobtrusively sit, quietly monitoring and tallying key-presses and mouse movements. At the end of a year, the customer would be asked to send in the disk file containing all the tallyings of the market research program. But again, most people would not bother to cooperate and some might see it as an invasion of privacy and of their disk space.

The perfect solution, from the company's point of view, would be a virus. Like any other virus it would be self-replicating and secretive. But it would not be destructive or facetious like an ordinary virus. Along with its self-replicating booster, it would contain a market research warhead. The virus would be released surreptitiously into the community of computer users. Just like an ordinary virus it would spread around, as people passed floppy disks and email around the community. As the virus spread from computer to computer, it would build up statistics on user behaviour, monitored secretly from deep within a succession of systems. Every now and again, a copy of the virus would happen to find its way, by normal epidemic traffic, back into one of the company's own computers. There it would be debriefed and its data collated with data from other copies of the virus that had come 'home'.

Looking into the future, it is not fanciful to imagine a time when viruses, both bad and good, have become so ubiquitous that we could speak of an ecological community of viruses and legitimate programs coexisting in the silicosphere. At present, software is advertised as, say, 'Compatible with System 7'. In the future, products may be advertised as 'Compatible with all viruses registered in the 2008 World Virus Census; immune to all listed virulent viruses; takes full advantage of the facilities offered by the following benign viruses if present ...' Word-processing software, say, may hand over particular functions, such as word-counting and string-searches, to friendly viruses burrowing autonomously through the text.

Looking even further into the future, whole integrated software systems might grow, not by design, but by something like the growth of an ecological community such as a tropical rainforest. Gangs of mutually compatible viruses might grow up, in the same way as genomes can be regarded as gangs of mutually compatible genes. Indeed, I have even suggested that our genomes should be regarded as gigantic colonies of viruses. Genes cooperate with one another in genomes because natural selection has favoured those genes that prosper in the presence of the other genes that happen to be common in the gene pool. Different gene pools may evolve towards different combinations of mutually compatible genes. I envisage a time when, in the same kind of way, computer viruses may evolve towards compatibility with other viruses, to form communities or gangs. But then again, perhaps not! At any rate, I find the speculation more alarming than exciting.

At present, computer viruses don't strictly evolve. They are invented by human programmers and if they evolve they do so in the same weak sense as cars or aeroplanes evolve. Designers derive this year's car as a slight modification of last year's car, and they may, more or less consciously, continue a trend of the last few years – further flattening of the radiator grill or whatever it may be. Computer virus designers dream up ever more devious tricks for outwitting the programmers of antivirus software. But computer viruses don't – so far – mutate and evolve by true natural selection. They may do so in the future. Whether they evolve by natural selection, or whether their evolution is steered by human designers, may not make much difference to their eventual performance. By either kind of evolution, we expect them to become better at concealment, and we expect them to become subtly compatible with other viruses that are at the same time prospering in the computer community.

DNA viruses and computer viruses spread for the same reason: an environment exists in which there is machinery well set up to duplicate and spread them around and to obey the instructions that the viruses embody. These two environments are, respectively, the environment of cellular physiology and the environment provided by a large community of computers and data-handling machinery. Are there any other environments like these, any other humming paradises of replication?

The Infected Mind

I have already alluded to the programmed-in gullibility of a child, so useful for learning language and traditional wisdom, and so easily subverted by nuns, Moonies and their ilk. More generally, we all exchange information with one another. We don't exactly plug floppy disks into slots in one another's skulls, but we exchange sentences, both through our ears and through our eyes. We notice each other's styles of moving and of dressing, and are influenced. We take in advertising jingles, and are presumably persuaded by them, otherwise hard-headed businessmen would not spend so much money polluting the air with them.

Think about the two qualities that a virus, or any sort of parasitic replicator, demands of a friendly medium: the two qualities that make cellular machinery so friendly towards parasitic DNA, and that make computers so friendly towards computer viruses. These qualities are, first, a readiness to replicate information accurately, perhaps with some mistakes that are subsequently reproduced accurately; and, second, a readiness to obey instructions encoded in the information so replicated. Cellular machinery and electronic computers excel in both these virus-friendly qualities. How do human brains match up? As faithful duplicators they are certainly less perfect than either cells or electronic computers. Nevertheless, they are still pretty good, perhaps about as faithful as an RNA virus, though not as good as DNA with all its elaborate proofreading measures against textual degradation. Evidence of the fidelity of brains, especially child brains, as data duplicators, is provided by language itself. Bernard Shaw's Professor Higgins was able by ear alone to place Londoners in the street where they grew up. Fiction is not evidence for anything, but everyone knows that Higgins's fictional skill is only an exaggeration of something we can all do. Any American can tell Deep South from Mid West, New England from Hillbilly. Any New Yorker can tell Bronx from Brooklyn. Equivalent claims could be substantiated for any country. What this phenomenon means is that human brains are capable of pretty accurate copying (otherwise the accents of, say, Newcastle would not be stable enough to be recognized) but with some mistakes (otherwise pronunciation would not evolve, and all speakers of a language would inherit identically the same accents from their remote ancestors). Language evolves, because it has both the great stability and the slight changeability that are prerequisites for any evolving system.

The second requirement of a virus-friendly environment – that it should obey a program of coded instructions – is again only quantitatively less true for brains than for cells or computers. We sometimes obey orders from one another, but also we sometimes don't. Nevertheless, it is a telling fact that, the world over, the vast majority of children follow the religion of their parents rather than any of the other available religions. Instructions to genuflect, to bow towards Mecca, to nod one's head rhythmically towards the wall, to shake like a maniac, to 'speak in tongues' – the list of such arbitrary and pointless motor patterns offered by religion alone is extensive – are obeyed, if not slavishly, at least with some reasonably high statistical probability.

Less portentously, and again especially prominent in children, the 'craze' is a striking example of behaviour that owes more to epidemiology than to rational choice. Yo-yos, hula hoops and pogo sticks, with their associated behavioural fixed actions, sweep through schools, and more sporadically leap from school to school, in patterns that differ from a measles epidemic in no serious particular. Ten years ago, you could have travelled thousands of miles through the United States and never seen a baseball cap turned back to front. Today the reverse baseball cap is ubiquitous. I do not know what the pattern of geographic spread of the reverse baseball cap precisely was, but epidemiology is certainly among the professions primarily qualified to study it. We don't have to get into arguments about 'determinism'; we don't have to claim that children are compelled to imitate their fellows' hat fashions. It is enough that their hat-wearing behaviour, as a matter of fact, *is* statistically affected by the hat-wearing behaviour of their fellows.

Trivial though they are, crazes provide us with yet more circumstantial evidence that human minds, especially perhaps juvenile ones, have the qualities that we have singled out as desirable for an informational parasite. At the very least the mind is a plausible *candidate* for infection by something like a computer virus, even if it is not quite such a parasite's dream-environment as a cell nucleus or an electronic computer. It is intriguing to wonder what it might feel like, from the inside, if one's mind were the victim of a 'virus'. This might be a deliberately designed parasite, like a present-day computer virus. Or it might be an inadvertently mutated and unconsciously evolved parasite. Either way, especially if the evolved parasite was the memetic descendant of a long line of successful ancestors, we are entitled to expect the typical 'mind virus' to be pretty good at its job of getting itself successfully replicated.

Progressive evolution of more effective mind-parasites will have two aspects. New 'mutants' (either random or designed by humans) that are better at spreading will become more numerous. And there will be a ganging up of ideas that flourish in one another's presence, ideas that mutually support one another just as genes do and, as I have speculated, computer viruses may one day do. We expect that replicators will go around together from brain to brain in mutually compatible gangs. These gangs will come to constitute a package, which may be sufficiently stable to deserve a collective name such as Roman Catholicism or Voodoo. It doesn't matter too much whether we analogize the whole package to a single virus, or each one of the component parts to a single virus. The analogy is not that precise anyway, just as the distinction between a computer virus and a computer worm is nothing to get worked up about. What matters is that minds are friendly environments to parasitic, self-replicating ideas or information, and that minds are typically massively infected.

Like computer viruses, successful mind viruses will tend to be hard for their victims to detect. If you are the victim of one, the chances are that you won't know it, and may even vigorously deny it. Accepting that a virus might be difficult to detect in your own mind, what tell-tale signs might you look out for? I shall answer by imagining how a medical textbook might describe the typical symptoms of a sufferer (arbitrarily assumed to be male).

1. The patient typically finds himself impelled by some deep, inner conviction that something is true, or right, or virtuous: a conviction that doesn't seem to owe anything to evidence or reason, but which, nevertheless, he feels as totally compelling and convincing. We doctors refer to such a belief as 'faith'.
2. Patients typically make a positive virtue of faith's being strong and unshakeable, *in spite of* not being based upon evidence. Indeed, they may feel that the less evidence there is, the more virtuous the belief (see below). This paradoxical idea that lack of evidence is a positive virtue where faith is concerned has

something of the quality of a program that is self-sustaining, because it is self-referential.* Once the proposition is believed, it automatically undermines opposition to itself. The 'lack of evidence is a virtue' idea would be an admirable sidekick, ganging up with faith itself in a clique of mutually supportive viral programs.

3.A related symptom, which a faith-sufferer may also present, is the conviction that 'mystery', per se, is a good thing. It is not a virtue to solve mysteries. Rather we should enjoy them, even revel in their insolubility.

Any impulse to solve mysteries could be seriously inimical to the spread of a mind virus. It would not, therefore, be surprising if the idea that 'mysteries are better not solved' was a favoured member of a mutually supporting gang of viruses. Take the 'Mystery of the Transubstantiation'. It is easy and non-mysterious to believe that in some symbolic or metaphorical sense the eucharistic wine turns into the blood of Christ. The Roman Catholic doctrine of transubstantiation, however, claims far more. The 'whole substance' of the wine is converted into the blood of Christ; the appearance of wine that remains is 'merely accidental', 'inhering in no substance'. Transubstantiation is colloquially taught as meaning that the wine 'literally' turns into the blood of Christ. Whether in its obfuscatory Aristotelian or its franker colloquial form, the claim of transubstantiation can be made only if we do serious violence to the normal meanings of words like 'substance' and 'literally'. Redefining words is not a sin but, if we use words like 'whole substance' and 'literally' for this case, what word are we going to use when we really and truly *want* to say that something did actually happen? As Anthony Kenny observed of his own puzzlement as a young seminarian, 'For all I could tell, my typewriter might be Benjamin Disraeli transubstantiated ...'

Roman Catholics, whose belief in infallible authority compels them to accept that wine becomes physically transformed into blood despite all appearances, refer to the 'Mystery' of the transubstantiation. Calling it a Mystery makes everything OK, you

*This is among many related ideas that have been grown in the endlessly fertile mind of Douglas Hofstadter (*Metamagical Themas*, London, Penguin, 1985).

see. At least, it works for a mind well prepared by background infection. Exactly the same trick is performed in the 'Mystery' of the Trinity. Mysteries are not meant to be solved, they are meant to strike awe. The 'mystery is a virtue' idea comes to the aid of the Catholic, who would otherwise find intolerable the obligation to believe the obvious nonsense of the transubstantiation and the 'three-in-one'. Again, the belief that 'mystery is a virtue' has a self-referential ring. As Douglas Hofstadter might put it, the very mysteriousness of the belief moves the believer to perpetuate the mystery.

An extreme symptom of 'mystery is a virtue' infection is Tertullian's '*Certum est quia impossibile est*' (It is certain because it is impossible). That way madness lies. One is tempted to quote Lewis Carroll's White Queen, who, in response to Alice's 'One can't believe impossible things', retorted, 'I daresay you haven't had much practice ... When I was your age, I always did it for half-an-hour a day. Why, sometimes I've believed as many as six impossible things before breakfast.' Or Douglas Adams's Electric Monk, a labour-saving device programmed to do your believing for you, which was capable of 'believing things they'd have difficulty believing in Salt Lake City' and which, at the moment of being introduced to the reader, believed, contrary to all the evidence, that everything in the world was a uniform shade of pink. But White Queens and Electric Monks become less funny when you realize that these virtuoso believers are indistinguishable from revered theologians in real life. 'It is by all means to be believed, because it is absurd' (Tertullian again). Sir Thomas Browne quotes Tertullian with approval, and goes further: 'Methinks there be not impossibilities enough in religion for an active faith.' And 'I desire to exercise my faith in the difficultest point; for to credit ordinary and visible objects is not faith, but perswasion.'[80] I have the feeling that something more interesting is going on here than just plain insanity or surrealist nonsense, something akin to the admiration we feel when we watch a juggler on a tightrope. It is as though the faithful gain prestige through managing to believe even more ridiculous things than their rivals succeed in believing. Are these people testing – exercising – their believing muscles, training themselves to believe impossible things so that they can take in their stride the merely improbable things that they are ordinarily called upon to believe?

While I was writing this, The *Guardian* (29 July 1991) fortuitously carried a beautiful example. It came in an interview with a rabbi undertaking the bizarre task of vetting the kosher-purity of food products right back to the ultimate origins of their minutest ingredients. He was currently agonizing over whether to go all the way to China to scrutinize the menthol that goes into cough sweets.

> Have you ever tried checking Chinese menthol … it was extremely difficult, especially since the first letter we sent received the reply in best Chinese English, 'The product contains no kosher' … China has only recently started opening up to kosher investigators. The menthol should be OK, but you can never be absolutely sure unless you visit.

These kosher investigators run a telephone hotline on which up-to-the-minute red-alerts of suspicion are recorded against chocolate bars or cod-liver oil. The rabbi sighs that the green-inspired trend away from artificial colours and flavours 'makes life miserable in the kosher field because you have to follow all these things back'. When the interviewer asks him why he bothers with this obviously pointless exercise, he makes it very clear that the point is precisely that there *is* no point:

> That most of the Kashrut laws are divine ordinances without reason given is 100 per cent the point. It is very easy not to murder people. Very easy. It is a little bit harder not to steal because one is tempted occasionally. So that is no great proof that I believe in God or am fulfilling His will. But, if He tells me not to have a cup of coffee with milk in it with my mince-meat and peas at lunchtime, that is a test. The only reason I am doing that is because I have been told to so do. It is doing something difficult.

Helena Cronin has suggested to me that there may be an analogy here to Amotz Zahavi's handicap theory of sexual selection and the evolution of signals.[81] Long unfashionable, even ridiculed, Zahavi's theory has recently been cleverly rehabilitated by Alan Grafen[82] and is now taken seriously by evolutionary biologists. Zahavi suggests that peacocks, for instance, evolve their absurdly burdensome fans with their ridiculously conspicuous (to predators) colours, precisely *because* they are burdensome and dangerous, and therefore impressive to females. The peacock is, in effect, saying: 'Look how fit and strong I must be, since I can afford to carry around this preposterous tail.'

To avoid misunderstanding of the subjective language in which Zahavi likes to make his points, I should add that the biologist's convention of personifying the unconscious actions of natural selection is taken for granted here. Grafen has translated the argument into an orthodox Darwinian mathematical model, and it works. No claim is being made here about the intentionality or awareness of peacocks and peahens. They can be as automatic or as intentional as you please. Moreover, Zahavi's theory is general enough not to depend upon a Darwinian underpinning. A flower advertising its nectar to a 'sceptical' bee could benefit from the Zahavi principle. But so could a human salesman seeking to impress a client.

The premise of Zahavi's idea is that natural selection will favour scepticism among females (or among recipients of advertising messages generally). The only way for a male (or any advertiser) to authenticate his boast of strength (quality, or whatever it is) is to prove that it is true by shouldering a truly costly handicap – a handicap *that only a genuinely strong* (high-quality, etc.) male could bear. It may be called the principle of costly authentication. And now to the point. Is it possible that some religious doctrines are favoured not *in spite of* being ridiculous but precisely *because* they are ridiculous? Any wimp in religion could believe that bread *symbolically* represents the body of Christ, but it takes a real, red-blooded Catholic to believe something as daft as the transub-stantiation. If you can believe that you can believe anything, and (witness the story of Doubting Thomas) these people are trained to see that as a virtue.

Let us return to our list of symptoms that someone afflicted with the mental virus of faith, and its accompanying gang of secondary infections, may expect to experience.

4. The sufferer may find himself behaving intolerantly towards vectors of rival faiths, in extreme cases even killing them or advocating their deaths. He may be similarly violent in his

disposition towards apostates (people who once held the faith but have renounced it); or towards heretics (people who espouse a different – often, perhaps significantly, only very slightly different – version of the faith). He may also feel hostile towards other modes of thought that are potentially inimical to his faith, such as the method of scientific reason which could function rather like a piece of antiviral software.

The threat to kill the distinguished novelist Salman Rushdie is only the latest in a long line of sad examples. On the very day that I wrote this, the Japanese translator of *The Satanic Verses* was found murdered, a week after a near-fatal attack on the Italian translator of the same book. By the way, the apparently opposite symptom of 'sympathy' for Muslim 'hurt', voiced by the Archbishop of Canterbury and other Christian leaders (verging, in the case of the Vatican, on outright criminal complicity) is, of course, a manifestation of the symptom we diagnosed earlier: the delusion that faith, however obnoxious its results, has to be respected simply because it *is* faith.

Murder is an extreme, of course. But there is an even more extreme symptom, and that is suicide in the militant service of a faith. Like a soldier ant programmed to sacrifice her life for germ-line copies of the genes that did the programming, a young Arab is taught that to die in a holy war is the quickest way to heaven. Whether the leaders who exploit him really believe this does not diminish the brutal power that the 'suicide mission virus' wields on behalf of the faith. Of course suicide, like murder, is a mixed blessing: would-be converts may be repelled by, or may treat with contempt, a faith that is insecure enough to need such tactics.

More obviously, if too many individuals sacrifice themselves the supply of believers could run low. This was true of a notorious example of faith-inspired suicide, though in this case it was not 'kamikaze' death in battle. The Peoples' Temple sect went extinct when its leader, the Reverend Jim Jones, led the bulk of his followers from the United States to the Promised Land of 'Jonestown' in the Guyanan jungle, where he persuaded more than 900 of them, children first, to drink cyanide. The macabre affair was fully investigated by a team from the *San Francisco Chronicle*.

> Jones, 'the Father', had called his flock together and told them it was time to depart for heaven.
> 'We're going to meet,' he promised, 'in another place.'
> The words kept coming over the camp's loudspeakers.
> 'There is great dignity in dying. It is a great demonstration for everyone to die.'[83]

Incidentally, it does not escape the trained mind of the alert sociobiologist that Jones, within his sect in earlier days, 'proclaimed himself the only person permitted to have sex' (presumably his partners were also permitted). A secretary would arrange for Jones's liaisons. She would call up and say, 'Father hates to do this, but he has this tremendous urge and could you please...?' His victims were not only female. One 17-year-old male follower, from the days when Jones's community was still in San Francisco, told how he was taken for dirty weekends to a hotel where Jones received a 'minister's discount for Rev. Jim Jones and son'. The same boy said:

> I was really in awe of him. He was more than a father. I would have killed my parents for him.

What is remarkable about the Reverend Jim Jones is not his own self-serving behaviour but the almost superhuman gullibility of his followers. Given such prodigious credulity, can anyone doubt that human minds are ripe for malignant infection?

Admittedly, the Reverend Jones conned only a few thousand people. But his case is an extreme, the tip of an iceberg. The same eagerness to be conned by religious leaders is widespread. Most of us would have been prepared to bet that nobody could get away with going on television and saying, in all but so many words, 'Send me your money, so that I can use it to persuade other suckers to send me their money too.' Yet today, in every major conurbation in the United States, you can find at least one television evangelist channel entirely devoted to this transparent confidence trick. And they get away with it in sackfuls. Faced with suckerdom on this awesome scale, it is hard not to feel a grudging sympathy with the shiny-suited conmen. Until you realize that not all the suckers are rich, and that it is often widows' mites on which the evangelists are growing fat. I have even heard one of them explicitly invoking the principle that I now identify with Zahavi's principle of costly authentication. God really appreciates a donation, he said with passionate sincerity, only when that donation is so large that it hurts. Elderly paupers were wheeled on to testify how much happier they felt since they had made over their little all to the Reverend whoever it was.

5. The patient may notice that the particular convictions that he holds, while having nothing to do with evidence, do seem to owe a great deal to epidemiology. Why, he may wonder, do I hold *this* set of convictions rather than *that* set? Is it because I surveyed all the world's faiths and chose the one whose claims seemed most convincing? Almost certainly not. If you have a faith, it is statistically overwhelmingly likely that it is the same faith as your parents and grandparents had. No doubt soaring cathedrals, stirring music, moving stories and parables help a bit. But by far the most important variable determining your religion is the accident of birth. The convictions that you so passionately believe would have been a completely different and largely contradictory set of convictions, if only you had happened to be born in a different place. Epidemiology, not evidence.

6. If the patient is one of the rare exceptions who follows a different religion from his parents, the explanation may still be epidemiological. To be sure, it is *possible* that he dispassionately surveyed the world's faiths and chose the most convincing one. But it is statistically more probable that he has been exposed to a particularly potent infective agent – a John Wesley, a Jim Jones or a St Paul. Here we are talking about horizontal transmission, as in measles. Before, the epidemiology was that of vertical transmission as in Huntington's Disease.

7. The internal sensations of the patient may be startlingly reminiscent of those more ordinarily associated with sexual love. This is an extremely potent force in the brain, and it is not surprising that some viruses have evolved to exploit it. St Teresa of Avila's famously orgasmic vision is too notorious to need quoting

again. More seriously, and on a less crudely sensual plane, the philosopher Anthony Kenny provides moving testimony to the pure delight that awaits those that manage to believe in the mystery of the transubstantiation. After describing his ordination as a Roman Catholic priest, empowered by laying on of hands to celebrate Mass, he vividly recalls

> … the exaltation of the first months during which I had the power to say Mass. Normally a slow and sluggish riser, I would leap early out of bed, fully awake and full of excitement at the thought of the momentous act I was privileged to perform. I rarely said the public Community Mass: most days I celebrated alone at a side altar with a junior member of the College to serve as acolyte and congregation. But that made no difference to the solemnity of the sacrifice or the validity of the consecration.
>
> It was touching the body of Christ, the closeness of the priest to Jesus, which most enthralled me. I would gaze on the Host after the words of consecration, soft-eyed like a lover looking into the eyes of his beloved … Those early days as a priest remain in my memory as days of fulfilment and tremulous happiness; something precious, and yet too fragile to last, like a romantic love-affair brought up short by the reality of an ill-assorted marriage.[84]

Dr Kenny is affectingly believable that it felt to him, as a young priest, as though he was in love with the consecrated host. What a brilliantly successful virus! On the same page, incidentally, Kenny also shows us that the virus is transmitted contagiously – if not literally, then at least in some sense – from the palm of the infecting bishop's hand through the top of the new priest's head:

> If Catholic doctrine is true, every priest validly ordained derives his orders in an unbroken line of laying on of hands, through the bishop who ordains him, back to one of the twelve Apostles … there must be centuries-long, recorded chains of layings on of hands. It surprises me that priests never seem to trouble to trace their spiritual ancestry in this way, finding out who ordained their bishop, and who ordained him, and so on to Julius II or Celestine V or Hildebrand, or Gregory the Great, perhaps.

It surprises me, too.

Is Science a Virus?

No. Not unless all computer programs are viruses. Good, useful programs spread because people evaluate them, recommend them and pass them on. Computer viruses spread solely because they embody the coded instructions: 'Spread me.' Scientific ideas, like all memes, are subject to a kind of natural selection, and this might look superficially virus-like, but the selective forces that scrutinize scientific ideas are not arbitrary or capricious. They are exacting, well-honed rules, and they do not favour pointless self-serving behaviour. They favour all the virtues laid out in textbooks of standard methodology: testability, evidential support, precision, quantifiability, consistency, intersubjectivity, repeatability, universality, progressiveness, independence of cultural milieu, and so on. Faith spreads despite a total lack of every single one of these virtues.

You may find elements of epidemiology in the spread of scientific ideas, but it will be largely descriptive epidemiology. The rapid spread of a good idea through the scientific community may even look like a description of a measles epidemic. But when you examine the underlying reasons you find that they are good ones, satisfying the demanding standards of scientific method. In the history of the spread of faith you will find little else but epidemiology, and causal epidemiology at that. The reason why person A believes one thing and B believes another is simply and solely that A was born on one continent and B on another. Testability, evidential support and the rest aren't even remotely considered. For scientific belief, epidemiology merely comes along afterwards and describes the history of its acceptance. For religious belief, epidemiology is the root cause.

Epilogue

Happily, viruses don't win every time. Many children emerge unscathed from the worst that nuns and mullahs can throw at them. Anthony Kenny's own story has a happy ending. He eventually renounced his orders because he could no longer tolerate the obvious contradictions within Catholic belief, and he is now a highly respected scholar. But one cannot help remarking that it must be a powerful infection indeed that took a man of his wisdom and intelligence – now President of the British Academy, no less – three decades to fight off. Am I unduly alarmist to fear for the soul of my six-year-old innocent?

The Great Convergence[85]

Are science and religion converging? No. There *are* modern scientists whose words sound religious but whose beliefs, on close examination, turn out to be identical to those of other scientists who straightforwardly call themselves atheists. Ursula Goodenough's lyrical book, *The Sacred Depths of Nature*,[86] is sold as a religious book, is endorsed by theologians on the back cover, and its chapters are liberally laced with prayers and devotional meditations. Yet, by the book's own account, Dr Goodenough does not believe in any sort of supreme being, does not believe in any sort of life after death; on any normal understanding of the English language, she is no more religious than I am. She shares with other atheist scientists a feeling of awe at the majesty of the universe and the intricate complexity of life. Indeed, the jacket copy for her book – the message that science does not 'point to an existence that is bleak, devoid of meaning, pointless ...' but on the contrary 'can be a wellspring of solace and hope' – would have been equally suitable for my own *Unweaving the Rainbow*, or Carl Sagan's *Pale Blue Dot*.[87] If that is religion, then I am a deeply religious man. But it isn't. As far as I can tell, my 'atheistic' views are identical to Ursula Goodenough's 'religious' ones. One of us is misusing the English language, and I don't think it's me.

She happens to be a biologist but this kind of neo-deistic pseudo-religion is more often associated with physicists. In Stephen Hawking's case, I hasten to insist, the accusation is unjust. His much quoted phrase 'The Mind of God' no more indicates belief in God than does my 'God knows!' (as a way of saying that I don't). I suspect the same of Einstein's picturesque invoking of the 'Dear Lord' to personify the laws of physics*. Paul Davies, however, adopted Hawking's phrase as the title of a book which went on to earn the Templeton Prize for Progress in Religion, the most lucrative prize in the world today, prestigious enough to be presented in Westminster Abbey by royalty. Daniel Dennett once remarked to me in Faustian vein: 'Richard, if ever you fall on hard times ...'

The latter day deists have moved on from their eighteenth-century counterparts who, for all that they eschewed revelation and espoused no particular denomination, still believed in some sort of supreme intelligence. If you count Einstein and Hawking as religious, if you allow the cosmic awe of Ursula Goodenough, Paul Davies, Carl Sagan and me as true religion, then religion and science have indeed converged, especially when you factor in such atheist priests as Don Cupitt and many university chaplains. But if 'religion' is allowed such a flabbily elastic definition, what word is left for *real* religion, religion as the ordinary person in the pew or on the prayer-mat understands it today; religion, indeed, as any intellectual would have understood it in previous centuries, when intellectuals were religious like everybody else? If God is a synonym for the deepest principles of physics, what word is left for a hypothetical being who answers prayers; intervenes to save cancer patients or help evolution over difficult jumps; forgives sins or dies for them? If we are allowed to relabel scientific awe as a religious impulse, the case goes through on the nod. You have *redefined* science as religion, so it's hardly surprising if they turn out to 'converge'.

---

*Indeed, Einstein himself was indignant at the suggestion: 'It was, of course, a lie what you read about my religious convictions, a lie which is being systematically repeated. I do not believe in a personal God and I have never denied this but have expressed it clearly. If something is in me which can be called religious then it is the unbounded admiration for the structure of the world so far as our science can reveal it.' From Albert Einstein, *The Human Side*, ed. H. Dukas and B. Hoffman (Princeton, Princeton University Press, 1981). The lie is still being systematically spread about, carried through the meme pool by the desperate desire so many people have to believe it – such is Einstein's prestige.

Another kind of convergence has been alleged between modern physics and eastern mysticism. The argument goes essentially as follows. Quantum mechanics, that brilliantly successful flagship theory of modern science, is deeply mysterious and hard to understand. Eastern mystics have always been deeply mysterious and hard to understand. Therefore eastern mystics must have been talking about quantum theory all along. Similar mileage is made of Heisenberg's Uncertainty Principle ('Aren't we all, in a very real sense, uncertain?'), Fuzzy Logic ('Yes, it's OK for you to be fuzzy too'), Chaos and Complexity Theory (the butterfly effect, the platonic, hidden beauty of the Mandelbrot Set – you name it, somebody has mysticized it and turned it into dollars). You can buy any number of books on 'quantum healing', not to mention quantum psychology, quantum responsibility, quantum morality, quantum aesthetics, quantum immortality and quantum theology. I haven't found a book on quantum feminism, quantum financial management or Afro-quantum theory, but give it time. The whole dippy business is ably exposed by the physicist Victor Stenger in his book *The Unconscious Quantum*, from which the following gem is taken.[88] In a lecture on 'Afrocentric healing', the psychiatrist Patricia Newton said that traditional healers

> ... are able to tap that other realm of negative entropy – that super-quantum velocity and frequency of electromagnetic energy and bring them as conduits down to our level. It's not magic. It's not mumbo-jumbo. You will see the dawn of the twenty-first century, the new medical quantum physics really distributing these energies and what they are doing.

Sorry, mumbo-jumbo is precisely what it is. Not African mumbo-jumbo but pseudoscientific mumbo-jumbo, even down to the trademark misuse of 'energy'. It is also religion, masquerading as science in a cloying love-feast of bogus convergence.

In 1996 the Vatican, fresh from its magnanimous reconciliation with Galileo a mere 350 years after his death, publicly announced that evolution had been promoted from tentative hypothesis to accepted theory of science*. This is less dramatic than many American Protestants think it is, for the Roman Church, whatever its faults, has never been noted for biblical literalism – on the contrary, it has treated the Bible with suspicion, as something close to a subversive document, needing to be carefully filtered through priests rather than given raw to congregations. The Pope's recent message on evolution has, nevertheless, been hailed as another example of late twentieth-century convergence between science and religion. Responses to the Pope's message exhibited liberal intellectuals at their worst, falling over themselves in their agnostic eagerness to concede to religion its own 'magisterium'†, of equal importance to that of science, but not opposed to it, not even overlapping it. Such agnostic conciliation is, once again, easy to mistake for genuine convergence, a true meeting of minds.

At its most naive, this intellectual appeasement policy partitions up the intellectual territory into 'how questions' (science) and 'why questions' (religion). What *are* 'why questions', and why should we feel entitled to think they deserve an answer?

---

*This is to give the Pope the benefit of the doubt. The key passage in the original French version of his message is, *'Aujourd'hui ... de nouvelles connaissances conduisent à reconnaître dans la théorie de l'évolution plus qu'une hypothèse.'* The official English translation rendered *'plus qu'une hypothèse'* as 'more than one hypothesis'. *'Une'* is ambiguous in French, and it has been charitably suggested that what the Pope really meant was that evolution is 'more than a [mere] hypothesis'. If the official English version is indeed a mis-translation, it is at best a spectacularly incompetent piece of work. It was certainly a godsend to opponents of evolution within the Catholic Church. The *Catholic World Report* eagerly seized upon 'more than one hypothesis' to conclude that there was a 'lack of unanimity within the scientific community itself'. The official Vatican line now favours the 'more than a mere hypothesis' interpretation, and this is fortunately how the news media have taken it. On the other hand, a later passage in the Pope's message seems consonant with the possibility that the official English translation got it right after all: 'And, to tell the truth, rather than *the* theory of evolution, we should speak of *several* theories of evolution.' Perhaps the Pope is simply confused, and doesn't know what he means.

†The word appears in a section heading, 'Evolution and the Church's Magisterium', in the official English version of the Pope's message, but not in the original French version, which has no section headings. Responses to the Pope's message, and the text of the message itself, including one by me, were published in the *Quarterly Review of Biology*, **72** (1992), 4.

There may be some deep questions about the cosmos that are forever beyond science. The mistake is to think that they are therefore not beyond religion too. I once asked a distinguished astronomer, a fellow of my college, to explain the Big Bang to me. He did so to the best of his (and my) ability, and I then asked what it was about the fundamental laws of physics that made the spontaneous origin of space and time possible. 'Ah,' he smiled, 'Now we move beyond the realm of science. This is where I have to hand over to our good friend the Chaplain.' But why the Chaplain? Why not the gardener or the chef? Of course chaplains, unlike chefs and gardeners, *claim* to have some insight into ultimate questions. But what reason have we ever been given for taking their claim seriously? Once again, I suspect that my friend the Professor of Astronomy was using the Einstein/Hawking trick of letting 'God' stand for 'That which we don't understand'. It would be a harmless trick if it were not continually misunderstood by those hungry to misunderstand it. In any case, optimists among scientists, of whom I am one, will insist that 'That which we don't understand' means only 'That which we don't *yet* understand'. Science is still working on the problem. We don't know where, or even whether, we shall ultimately be brought up short.

Agnostic conciliation, the decent liberal bending over backwards to concede as much as possible to anybody who shouts loudly enough, reaches ludicrous lengths in the following common piece of sloppy thinking. It goes roughly like this. You can't prove a negative (so far so good). Science has no way to disprove the existence of a supreme being (this is strictly true). Therefore belief (or disbelief) in a supreme being is a matter of pure individual inclination, and they are therefore both equally deserving of respectful attention! When you say it like that the fallacy is almost self-evident: we hardly need spell out the *reductio ad absurdum*. To borrow a point from Bertrand Russell, we must be equally agnostic about the theory that there is a china teapot in elliptical orbit around the Sun. We can't disprove it. But that doesn't mean the theory that there is a teapot is on level terms with the theory that there isn't.

Now, if it be retorted that there actually are reasons X, Y and Z for finding a supreme being more plausible than a celestial teapot, then X, Y and Z should be spelled out because, if legitimate, they are proper scientific arguments which should be evaluated on their merits. Don't protect them from scrutiny behind a screen of agnostic tolerance. If religious arguments are actually better than Russell's teapot, let us hear the case. Otherwise, let those who call themselves agnostic with respect to religion add that they are equally agnostic about orbiting teapots. At the same time, modern theists might acknowledge that, when it comes to Baal and the Golden Calf, Thor and Wotan, Poseidon and Apollo, Mithras and Ammon Ra, they are actually atheists. We are all atheists about most of the gods that humanity has ever believed in. Some of us just go one god further.

In any case, the belief that religion and science occupy separate magisteria is dishonest.[89] It founders on the undeniable fact that religions still make claims about the world which, on analysis, turn out to be scientific claims. Moreover, religious apologists try to have it both ways, to eat their cake and have it. When talking to intellectuals, they carefully keep off science's turf, safe inside the separate and invulnerable religious magisterium. But when talking to a non-intellectual mass audience they make wanton use of miracle stories, which are blatant intrusions into scientific territory. The Virgin Birth, the Resurrection, the Raising of Lazarus, the manifestations of Mary and the Saints around the Catholic world, even the Old Testament miracles, all are freely used for religious propaganda, and very effective they are with an audience of unsophisticates and children. Every one of these miracles amounts to a scientific claim, a violation of the normal running of the natural world. Theologians, if they want to remain honest, should make a choice. You can claim your own magisterium, separate from science's but still deserving of respect. But in that case you have to renounce miracles. Or you can keep your Lourdes and your miracles, and enjoy their huge recruiting potential among the uneducated. But then you must kiss goodbye to separate magisteria and your high-minded aspiration to converge on science.

The desire to have it both ways is not surprising in a good propagandist. What is surprising is the readiness of liberal agnostics to go along with it; and their readiness to write off, as simplistic, insensitive extremists, those of us with the temerity to blow the whistle. The whistle-blowers are accused of flogging a dead horse, of imagining an outdated caricature of religion in which God has a long white beard and lives in a physical place called Heaven. Nowadays, we are told, religion has moved on. Heaven is not a physical place, and God does not have a physical body where a beard might sit. Well, yes, admirable: separate magisteria, real convergence. But the doctrine of the Assumption was defined as an Article of Faith by Pope Pius XII as recently as 1 November 1950, and is binding on all Catholics. It clearly states that the *body* of Mary was taken into Heaven and reunited with her soul. What can that mean, if not that Heaven is a physical place, physical enough to contain bodies? To repeat, this is not some quaint and obsolete tradition, with nowadays a purely symbolic significance. It was in the twentieth century that (to quote the 1996 *Catholic Encyclopedia*) 'Pope Pius XII declared infallibly that the Assumption of the Blessed Virgin Mary was a dogma of the Catholic Faith', thereby upgrading to the status of official dogma what his predecessor, Benedict XIV, also in the twentieth century, had called 'a probable opinion, which to deny were impious and blasphemous'.

Convergence? Only when it suits. To an honest judge, the alleged convergence between religion and science is a shallow, empty, hollow, spin-doctored sham.

Dolly and the Cloth Heads[90]

A news story like the birth of the cloned sheep Dolly is always followed by a flurry of energetic press activity. Newspaper columnists sound off, solemnly or facetiously; occasionally intelligently. Radio and television producers seize the telephone and round up panels to discuss and debate the moral and legal issues. Some of these panellists are experts on the science, as you would expect and as is right and proper. Equally appropriate are scholars of moral or legal philosophy. Both categories are invited to the studio in their own right, because of their specialized knowledge or their proven ability to think intelligently and speak clearly. The arguments that they have with each other are usually illuminating and rewarding.

The same cannot be said of the third, and most obligatory, category of studio guest: the religious lobby. Lobbies in the plural, I should say, because all the religions have to be represented. This incidentally multiplies the sheer number of people in the studio, with consequent consumption, if not waste, of time.

Out of good manners I shall not mention names, but during the admirable Dolly's week of fame I took part in broadcast or televised discussions of cloning with several prominent religious leaders, and it was not edifying. One of the most eminent of these spokesmen, recently elevated to the House of Lords, got off to a flying start by refusing to shake hands with the women in the television studio, apparently for fear they might be menstruating or otherwise 'unclean'. They took the insult more graciously than I would have, and with the 'respect' always bestowed on religious prejudice – but no other kind of prejudice. When the panel discussion got going, the woman in the chair, treating this bearded patriarch with great deference, asked him to spell out the harm that cloning might do, and he answered that atomic bombs were harmful. Yes indeed, no possibility of disagreement there. But wasn't the discussion supposed to be about cloning?

Since it was his choice to shift the discussion to atomic bombs, perhaps he knew more about physics than about biology? But no, having delivered himself of the daring falsehood that Einstein split the atom, the sage switched with confidence to history. He made the telling point that, since God laboured six days and then rested on the seventh, scientists too ought to know when to call a halt. Now, either he really believed that the world was made in six days, in which case his ignorance alone disqualifies him from being taken seriously, or, as the chairwoman charitably suggested, he intended the point purely as an allegory – in which case it was a lousy allegory. Sometimes in life it is a good idea to stop, sometimes it is a good idea to go on. The trick is to decide *when* to stop. The allegory of God resting on the seventh day cannot, in itself, tell us whether we have reached the right point to stop in some particular case. As allegory, the six-day creation story is empty. As history, it is false. So why bring it up?

The representative of a rival religion on the same panel was frankly confused. He voiced the common fear that a human clone would lack individuality. It would not be a whole, separate human being but a mere soulless automaton. When I warned him that his words might be offensive to identical twins, he said that identical twins were a quite different case. Why?

On a different panel, this time for radio, yet another religious leader was similarly perplexed by identical twins. He too had 'theological' grounds for fearing that a clone would not be a separate individual and would therefore lack 'dignity'. He was swiftly informed of the undisputed scientific fact that identical twins are clones of each other with the same genes, like Dolly except that Dolly is the clone of an older sheep. Did he really mean to say that identical twins (and we all know some) lack the dignity of separate individuality? His reason for denying the relevance of the twin analogy was very odd indeed. He had great faith, he informed us, in the power of nurture over nature. Nurture is why identical twins are really different individuals. When you get to know a pair of twins, he concluded triumphantly, they even *look* a bit different.

Er, quite so. And if a pair of clones were separated by fifty years, wouldn't their respective nurtures be even *more* different? Haven't you just shot yourself in your theological foot? He just didn't get it – but after all he hadn't been chosen for his ability to follow an argument. I don't want to sound uncharitable, but I submit to radio and television producers that merely being a spokesman for a particular 'tradition', 'faith' or 'community' may not be enough. Isn't a certain minimal qualification in the IQ department desirable too?

Religious lobbies, spokesmen of 'traditions' and 'communities', enjoy privileged access not only to the media but to influential committees of the great and the good, to governments and school boards. Their views are regularly sought, and heard with exaggerated 'respect', by parliamentary committees. You can be sure that, when an Advisory Commission is set up to advise on cloning policy, or any other aspect of reproductive technology, religious lobbies will be prominently represented. Religious spokesmen and spokeswomen enjoy an inside track to influence and power which others have to earn through their own ability or expertise. What is the justification for this?

Why has our society so meekly acquiesced in the convenient fiction that religious views have some sort of right to be respected automatically and without question? If I want you to respect my views on politics, science or art, I have to earn that respect by argument, reason, eloquence or relevant knowledge. I have to withstand counter-arguments. But if I have a view that is part of my religion, critics must respectfully tiptoe away or brave the indignation of society at large. Why are religious opinions off limits in this way? Why do we have to respect them, simply because they are religious?

How, moreover, do you decide which of many mutually contradictory religions should be granted this unquestioned respect: this unearned influence. If we invite a Christian spokesman into the television studio or the Advisory Committee, should it be a Catholic or a Protestant, or do we have to have both to make it fair? (In Northern Ireland the difference is, after all, important enough to constitute a recognized motive for murder.) If we have a Jew and a Muslim, must we have both Orthodox and Reformed, both Shiite and Sunni? And why not Moonies, Scientologists and Druids?

Society, for no reason that I can discern, accepts that parents have an automatic right to bring their children up with particular religious opinions and can withdraw them from, say, biology classes that teach evolution. Yet we'd all be scandalized if children were withdrawn from Art History classes that teach the merits of artists not to their parents' taste. We meekly agree, if a student says, 'Because of my religion I can't take my final examination on the day appointed so, no matter what the inconvenience, you'll have to set a special examination for me.' It is not obvious why we treat such a demand with any more respect than, say, 'Because of my basketball match (or because of my mother's birthday) I can't take the examination on a particular day.' Such favoured treatment for religious opinion reaches its apogee in wartime. A highly intelligent and sincere individual who justifies his personal pacifism by deeply thought-out moral philosophic arguments finds it hard to achieve Conscientious Objector status. If only he had been born into a religion whose scriptures forbid fighting, he'd have needed no other arguments at all. It is the same unquestioned respect for religions that causes society to beat a path to their leaders' doors whenever an issue like cloning is in the air. Perhaps, instead, we should listen to those whose words themselves justify our heeding them.

3.5
Time to Stand Up[91]

'To blame Islam for what happened in New York is like blaming Christianity for the troubles in Northern Ireland!'* Yes. Precisely. It is time to stop pussyfooting around. Time to get angry. And not only with Islam.

Those of us who have renounced one or another of the three 'great' monotheistic religions have, until now, moderated our language for reasons of politeness. Christians, Jews and Muslims are sincere in their beliefs and in what they find holy. We have respected that, even as we have disagreed with it. The late Douglas Adams put it with his customary good humour, in an impromptu speech in 1998[92] (slightly abridged):

> Now, the invention of the scientific method is, I'm sure we'll all agree, the most powerful intellectual idea, the most powerful framework for thinking and investigating and understanding and challenging the world around us that there is, and it rests on the premise that any idea is there to be attacked. If it withstands the attack then it lives to fight another day, and if it doesn't withstand the attack then down it goes. Religion doesn't seem to work like that. It has certain ideas at the heart of it which we call sacred or holy or whatever. What it means is, 'Here is an idea or a notion that you're not allowed to say anything bad about; you're just not. Why not? – because you're not!' If somebody votes for a party that you don't agree with, you're free to argue about it as much as you like; everybody will have an argument but nobody feels aggrieved by it. If somebody thinks taxes should go up or down, you are free to have an argument about it. But on the other hand if somebody says, 'I mustn't move a light switch on a Saturday', you say, 'I *respect* that.'

*Tony Blair is among many who have said something like this, thinking, wrongly, that to blame Christianity for Northern Ireland is self-evidently absurd.

The odd thing is, even as I am saying that I am thinking, 'Is there an Orthodox Jew here who is going to be offended by the fact that I just said that?' But I wouldn't have thought, 'Maybe there's somebody from the left wing or somebody from the right wing or somebody who subscribes to this view or the other in economics' when I was making the other points. I just think, 'Fine, we have different opinions'. But the moment I say something that has something to do with somebody's (I'm going to stick my neck out here and say irrational) beliefs, then we all become terribly protective and terribly defensive and say, 'No, we don't attack that; that's an irrational belief but no, we respect it.'

Why should it be that it's perfectly legitimate to support the Labour party or the Conservative party, Republicans or Democrats, this model of economics versus that, Macintosh instead of Windows – but to have an opinion about how the Universe began, about who created the Universe … no, that's holy? What does that mean? Why do we ring-fence that for any other reason other than that we've just got used to doing so? There's no other reason at all, it's just one of those things that crept into being and once that loop gets going it's very, very powerful. So, we are used to not challenging religious ideas, but it's very interesting how much of a furore Richard creates when he does it! Everybody gets absolutely frantic about it because you're not allowed to say these things. Yet when you look at it rationally there is no reason why those ideas shouldn't be as open to debate as any other, except that we have agreed somehow between us that they shouldn't be.

Douglas is dead, but his words are an inspiration to us now to stand up and break this absurd taboo.[93] My last vestige of 'hands off religion' respect disappeared in the smoke and choking dust of September 11th 2001, followed by the 'National Day of Prayer', when prelates and pastors did their tremulous Martin Luther King impersonation and urged people of mutually incompatible faiths to hold hands, united in homage to the very force that caused the problem in the first place. It is time for people of intellect, as opposed to people of faith, to stand up and say 'Enough!' Let our tribute to the September dead be a new resolve: to respect people for what they individually think, rather than respect groups for what they were collectively brought up to believe.

Notwithstanding bitter sectarian hatreds over the centuries (all too obviously still going strong), Judaism, Islam and Christianity have much in common. Despite New Testament watering down and other reformist tendencies, all three pay historical allegiance to the same violent and vindictive God of Battles, memorably summed up by Gore Vidal in 1998:

> The great unmentionable evil at the center of our culture is monotheism. From a barbaric Bronze Age text known as the Old Testament, three anti-human religions have evolved – Judaism, Christianity, and Islam. These are sky-god religions. They are, literally, patriarchal – God is the Omnipotent Father – hence the loathing of women for 2000 years in those countries afflicted by the sky-god and his earthly male delegates. The sky-god is a jealous god, of course. He requires total obedience from everyone on earth, as he is not just in place for one tribe, but for all creation. Those who would reject him must be converted or killed for their own good.

In The *Guardian* of 15 September 2001, I named belief in an afterlife as the key weapon that made the New York atrocity possible.[94] Of prior significance is religion's deep responsibility for the underlying hatreds that motivated people to use that weapon in the first place. To breathe such a suggestion, even with the most gentlemanly restraint, is to invite an onslaught of patronizing abuse, as Douglas Adams noted. But the insane cruelty of the suicide attacks, and the equally vicious, though numerically less catastrophic, 'revenge' attacks on hapless Muslims living in America and Britain, push me beyond ordinary caution.

How can I say that religion is to blame? Do I really imagine that, when a terrorist kills, he is motivated by a theological disagreement with his victim? Do I really think the Northern Ireland pub bomber says to himself, 'Take that, Tridentine Transubstantiationist bastards!' Of course I don't think anything of the kind. Theology is the last thing on the minds of such people. They are not killing because of religion, but because of political grievances, often justified. They are killing because the other lot killed their fathers. Or because the other lot drove their great grandfathers off their land. Or because the other lot oppressed our lot economically for centuries.

My point is not that religion itself is the motivation for wars, murders and terrorist attacks, but that religion is the principal *label*, and the most dangerous one, by which a 'they' as opposed to a 'we' can be identified at all. I am not even claiming that religion is the *only* label by which we identify the victims of our prejudice. There's also skin colour, language and social class. But often, as in Northern Ireland, these don't apply and religion is the only divisive label around. Even when it is not alone, religion is nearly always an incendiary ingredient in the mix as well. And please don't trot out Hitler as a counter-example. Hitler's sub-Wagnerian ravings constituted a religion of his own foundation, and his anti-Semitism owed a lot to his never-renounced Roman Catholicism.*

It is not an exaggeration to say that religion is the most inflammatory enemy-labelling device in history. Who killed your father? Not the individuals you are about to kill in 'revenge'. The culprits themselves have vanished over the border. The people who stole your great grandfather's land have died of old age. You aim your vendetta at those who belong to the same *religion* as the original perpetrators. It wasn't Seamus who killed your brother, but it was Catholics, so Seamus deserves to die 'in return'. Next, it was Protestants who killed Seamus so let's go out and kill some Protestants 'in revenge'. It was Muslims who destroyed the World

---

*'My feeling as a Christian points me to my Lord and Saviour as a fighter. It points me to the man who once in loneliness, surrounded by only a few followers, recognized these Jews for what they were and summoned men to the fight against them and who, God's Truth! was greatest not as sufferer but as fighter. In boundless love as a Christian and as a man I read through the passage which tells us how the Lord at last rose in His might and seized the scourge to drive out of the Temple the brood of vipers and of adders. How terrific was His fight for the world against the Jewish poison. Today, after two thousand years, with deepest emotion I recognize more profoundly than ever before – the fact that it was for this that He had to shed His blood upon the Cross. As a Christian I have no duty to allow myself to be cheated, but I have the duty to be a fighter for truth and justice. And as a man I have the duty to see to it that human society does not suffer the same catastrophic collapse as did the civilization of the ancient world some two thousand years ago – a civilization which was driven to its ruin through this same Jewish people.' Adolf Hitler, speech of 12 April 1922, Munich. From Norman H. Baynes (ed.), *The Speeches of Adolf Hitler, April 1922–August 1939* (2 vols., Oxford, Oxford University Press, 1942), vol. 1, pp. 19–20. See also http://www.secularhumanism.org/library/fi/murphy_19_2.html and http://www.nobeliefs.com/speeches.htm

Trade Center, so let's set upon the turbaned driver of a London taxi and leave him paralysed from the neck down.

The bitter hatreds that now poison Middle Eastern politics are rooted in the real or perceived wrong of the setting up of a Jewish State in an Islamic region. In view of all that the Jews had been through, it must have seemed a fair and humane solution. Probably deep familiarity with the Old Testament had given the European and American decision-makers some sort of idea that this really was the 'historic homeland' of the Jews (though the horrific biblical stories of how Joshua and others conquered their *Lebensraum* might have made them wonder). Even if it wasn't justifiable at the time, no doubt a good case can be made that, since Israel exists now, to try to reverse the *status quo* would be a worse wrong.

I do not intend to get into that argument. But if it had not been for religion, the very *concept* of a Jewish state would have had no meaning in the first place. Nor would the very concept of Islamic lands, as something to be invaded and desecrated. In a world without religion, there would have been no Crusades; no Inquisition; no anti-Semitic pogroms (the people of the diaspora would long ago have intermarried and become indistinguishable from their host populations); no Northern Ireland Troubles (no label by which to distinguish the two 'communities', and no sectarian schools to teach the children historic hatreds – they would simply be one community).

It is a spade we have here, let's *call* it a spade. The Emperor has no clothes. It is time to stop the mealy-mouthed euphemisms: 'Nationalists', 'Loyalists', 'Communities', 'Ethnic Groups', 'Cultures', 'Civilizations'. *Religions* is the word you need. Religions is the word you are struggling hypocritically to avoid.

Parenthetically, religion is unusual among divisive labels in being spectacularly *unnecessary*. If religious beliefs had any evidence going for them, we might have to accept them in spite of their concomitant unpleasantness. But there is no such evidence. To label people as death-deserving enemies because of disagreements about real world politics is bad enough. To do the same for disagreements about a delusional world inhabited by archangels, demons and imaginary friends is ludicrously tragic.

The resilience of this form of hereditary delusion is as astonishing as its lack of realism. It seems that control of the plane which crashed near Pittsburgh was probably wrestled out of the hands of the terrorists by a group of brave passengers. The wife of one of these valiant and heroic men, after she took the telephone call in which he announced their intention, said that God had placed her husband on the plane as His instrument to prevent the plane crashing into the White House. I have the greatest sympathy for this poor woman in her tragic loss, but just *think* about it! As my (also understandably overwrought) American correspondent who sent me this piece of news said:

> Couldn't God have just given the hijackers a heart attack or something instead of killing all those nice people on the plane? I guess he didn't give a flying fuck about the Trade Center, didn't bother to come up with a plan for them. [I apologize for my friend's intemperate language but, in the circumstances, who can blame her?]

Is there no catastrophe terrible enough to shake the faith of people, on both sides, in God's goodness and power? No glimmering realization that he might not be there at all: that we just might be on our own, needing to cope with the real world like grown-ups?

The United States is the most religiose country in Christendom, and its born-again leader is eyeball to eyeball with the most religiose people on Earth. Both sides believe that the Bronze Age God of Battles is on their side. Both take risks with the world's future in unshakeable, fundamentalist faith that God will grant them the victory. J. C. Squire's famous verse on the First World War spontaneously comes to mind:

> God heard the embattled nations sing and shout
> '*Gott strafe England*' and 'God save the King!'
> God this, God that, and God the other thing –
> 'Good God!' said God, 'I've got my work cut out!'

The human psyche has two great sicknesses: the urge to carry vendetta across generations, and the tendency to fasten group labels on people rather than see them as individuals. Abrahamic religion mixes explosively with (and gives strong sanction to) both. Only the wilfully blind could fail to implicate the divisive force of religion in most, if not all, of the violent enmities in the world today. Those of us who have for years politely concealed our contempt for the dangerous collective delusion of religion need to stand up and speak out. Things are different after September 11th. 'All is changed, changed utterly.'

4

THEY TOLD ME, HERACLITUS

One of the signs of growing older is that one ceases to be invited to be best man at weddings, or godfather at christenings. I have just begun to be called upon to write obituaries, speak eulogies and organize funerals. Jonathan Miller, on reaching the same landmark age, wrote a sad article, as an atheist, about atheist funerals. They are more than usually cheerless affairs, in his view. A funeral is the one occasion when he feels that religion actually has something to offer: not, of course, the delusion of an afterlife (as he would see it), but the hymns, the rituals, the vestments, the seventeenth-century words.

Loving the cadences of the Authorized Version and the Book of Common Prayer as I do, I surprise myself by the strength of my disagreement with Dr Miller. All funerals are sad, but secular funerals, properly organized, are hugely preferable on all counts. I have long noticed that even religious funerals are memorable mostly for their nonreligious content: the memoirs, the poems, the music. After listening to a well-crafted speech by someone who knew and loved the deceased, my feeling has been: 'Oh, it was so moving hearing so-and-so's tribute; if only there could have been more like that, and fewer of those empty, hollow prayers.' Secular funerals, by scrapping the prayers altogether, give more time for a beautiful memorial: a balance of tributes, music that evokes memories, poetry that may be alternately sad and uplifting, perhaps readings from the dead person's works, even some affectionate humour.

It is hard to think of the novelist Douglas Adams without affectionate humour, and it was much in evidence at his memorial service in the Church of St Martin in the Fields, in London. I was one of those who spoke, and my **eulogy** (4.2) is reprinted here, as the second piece in this section. But earlier – indeed, I finished it the day after he died – I wrote a **lament** (4.1) in The *Guardian*. The tone of these two pieces, one shocked and sad, the other affectionately celebratory, is so different that it seemed right to include both.

In the case of my revered colleague the evolutionary biologist W. D. Hamilton, it fell to me to organize his memorial service in the Chapel of New College, Oxford. I also spoke a **eulogy**, and it is reproduced as the third item (4.3) in this section. In this service, the music was provided by New College's wonderful choir. Two of the anthems had been sung at Darwin's funeral in Westminster Abbey, one of them specially composed for Darwin: a setting by Frederick Bridge of 'Happy is the man that findeth wisdom, and the man that getteth understanding' (Proverbs 3:13). I like to think that Bill, that dear, gentle, wise man, would have been pleased. At my suggestion the score has been reprinted in the posthumous volume of Bill's collected papers, *Narrow Roads of Gene Land*,[95] where it is certainly the only copy in print.

I met John Diamond only once, shortly before he died. I knew of him as a newspaper columnist and author of a courageous book, *C: Because cowards get cancer too*,[96] recounting his battle with a horrific form of throat cancer. When I met him at a cocktail party, he could not speak at all, and carried on lively and cheerful conversations by writing in a notebook. He was working on a second book, **Snake Oil** (4.4), taking the lid off the 'alternative' medicine which, while he was dying, was almost daily thrust his way by quacks or their well-meaning dupes. He died before he could complete the book, and I was honoured to be invited to write the Foreword for its posthumous publication.

Lament for Douglas[97]

This is not an obituary, there'll be time enough for them. It is not a tribute, not a considered assessment of a brilliant life, not a eulogy. It is a keening lament, written too soon to be balanced, too soon to be carefully thought through. Douglas, you cannot be dead.

A sunny Saturday morning in May, ten past seven, shuffle out of bed, log in to email as usual. The usual blue bold headings drop into place, mostly junk, some expected, and my gaze absently follows them down the page. The name Douglas Adams catches my eye and I smile. That one, at least, will be good for a laugh. Then I do the classic double-take, back up the screen. *What* did that heading actually say? **Douglas Adams died of a heart attack a few hours ago**. Then that other cliché, the words swelling before my eyes. It must be part of the joke. It must be some other Douglas Adams. This is too ridiculous to be true. I must still be asleep. I open the message, from a well-known German software designer. It is no joke, I am fully awake. And it is the right – or rather the wrong – Douglas Adams. A sudden heart attack, in the gym in Santa Barbara. 'Man, man, man, man oh man,'[k] the message concludes.

Man indeed, what a man. A giant of a man, surely nearer seven foot than six, broad-shouldered, and he did not stoop like some very tall men who feel uncomfortable with their height. But nor did he swagger with the macho assertiveness that can be intimidating in a big man. He neither apologized for his height, nor flaunted it. It was part of the joke against himself.

One of the great wits of our age, his sophisticated humour was founded in a deep, amalgamated knowledge of literature and science, two of my great loves. And he introduced me to my wife – at his fortieth birthday party. He was exactly her age, they had worked together on Dr Who. Should I tell her now, or let her sleep a bit longer before shattering her day? He initiated our togetherness and was a recurrently important part of it. I must tell her now.

Douglas and I met because I sent him an unsolicited fan letter – I think it is the only time I have ever written one. I had adored *The Hitchhiker's Guide to the Galaxy*. Then I read *Dirk Gently's Holistic Detective Agency*. As soon as I finished it I turned back to page one and read it straight through again – the only time I have ever done *that*, and I wrote to tell him so. He replied that he was a fan of my books, and he invited me to his house in London. I have seldom met a more congenial spirit. Obviously I knew he would be funny. What I didn't know was how deeply read he was in science. I should have guessed, for you can't understand many of the jokes in *Hitchhiker* if you don't know a lot of advanced science. And in modern electronic technology he was a real expert. We talked science a lot, in private, and even in public at literary festivals and on the wireless or television. And he became my guru on all technical problems. Rather than struggle with some ill-written and incomprehensible manual in Pacific Rim English, I would fire off an email to Douglas. He would reply, often within minutes, whether in London or Santa Barbara, or some hotel room anywhere in the world. Unlike most staffers of professional helplines, Douglas understood *exactly* my problem, knew *exactly* why it was troubling me, and always had the solution ready, lucidly and amusingly explained. Our frequent email exchanges brimmed with literary and scientific jokes and affectionately sardonic little asides. His technophilia shone through, but so did his rich sense of the absurd. The whole world was one big Monty Python sketch, and the follies of humanity are as comic in the world's silicon valleys as anywhere else.

He laughed at himself with equal good humour. At, for example, his epic bouts of writer's block ('I love deadlines. I love the whooshing noise they make as they go by') when, according to legend, his publisher and book agent would literally lock him in a hotel room, with no telephone, and nothing to do but write, releasing him only for supervised walks. If his enthusiasm ran away with him and he advanced a biological theory too eccentric for my professional scepticism to let pass, his mien at my dismissal of it would always be more humorously self-mocking than genuinely crestfallen. And he would have another go.

He laughed at his own jokes, which good comedians are not supposed to do, but he did it with such charm that the jokes became even funnier. He was gently able to poke fun without wounding, and it would be aimed not at individuals but at their absurd ideas. The moral of this parable, which he told with huge enjoyment, leaps out with no further explanation. A man didn't understand how televisions work, and was convinced that there must be lots of little men inside the box, manipulating images at high speed. An engineer explained to him about high frequency modulations of the electromagnetic spectrum, about transmitters and receivers, about amplifiers and cathode ray tubes, about scan lines moving across and down a phosphorescent screen. The man listened to the engineer with careful attention, nodding his head at every step of the argument. At the end he pronounced himself satisfied. He now really did understand how televisions work. 'But I expect there are just a *few* little men in there, aren't there?'

Science has lost a friend, literature has lost a luminary, the mountain gorilla and the black rhino have lost a gallant defender (he once climbed Kilimanjaro in a rhino suit to raise money to fight the cretinous trade in rhino horn), Apple Computer has lost its most eloquent apologist. And I have lost an irreplaceable intellectual companion and one of the kindest and funniest men I ever met. I officially received a happy piece of news yesterday, which would have delighted him. I wasn't allowed to tell anyone during the weeks I have secretly known about it, and now that I am allowed to it is too late.

The sun is shining, life must go on, seize the day and all those clichés. We shall plant a tree this very day: a Douglas Fir, tall, upright, evergreen. It is the wrong time of year, but we'll give it our best shot. Off to the arboretum.

The tree is planted, and this article completed, all within 24 hours of his death. Was it cathartic? No, but it was worth a try.

Eulogy for Douglas Adams

Church of St Martin in the Fields,
London, 17 September 2001

I believe it falls to me to say something about Douglas's love of science.* He once asked my advice. He was contemplating going back to university to read science, I think specifically my own subject of Zoology. I advised against it. He already knew plenty of science. It rings through almost every line he wrote and through the best jokes he made. As a single example, think of the Infinite Improbability Drive. Douglas thought like a scientist, but was much funnier. It is fair to say that he was a hero to scientists. And technologists, especially in the computer industry.

His unjustified humility in the presence of scientists came out touchingly in a magnificent impromptu speech at a Cambridge conference which I attended in 1998.[98] He was invited as a kind of honorary scientist – a thing that happened to him quite often. Thank goodness somebody switched on a tape recorder, and so we have the whole of this splendid extempore *tour de force*. It certainly ought to be published somewhere. I'm going to read a few disconnected paragraphs. He was a wonderful comedian as well as a brilliant comic writer, and you can hear his voice in every line:

> This was originally billed as a debate only because I was a bit anxious coming here ... in a room full of such luminaries, I thought, 'what could I, as an amateur, possibly have to say?' So I thought I would settle for a debate. But after having been here for a couple of days, I realised you're just a bunch of guys! ... I thought that what I'd do is stand up and have a debate with myself ... and hope sufficiently to provoke and inflame opinion that there'll be an outburst of chair-throwing at the end.
>
> Before I embark on what I want to try and tackle, may I warn you that things may get a little bit lost from time to time, because there's a lot of

*Others, of course, spoke of different aspects of his life.

stuff that's just come in from what we've been hearing today, so if I occasionally sort of go ... I have a four-year-old daughter and was very, very interested watching her face when she was in her first two or three weeks of life and suddenly realising what nobody would have realised in previous ages – she was rebooting!

I just want to mention one thing, which is completely meaningless, but I am terribly proud of – I was born in Cambridge in 1952 and my initials are DNA!

These inspired switches of subject are so characteristic of his style – and so endearing.

I remember once, a long time ago, needing a definition of life for a speech I was giving. Assuming there was a simple one and looking around the Internet, I was astonished at how diverse the definitions were and how very, very detailed each one had to be in order to include 'this' but not include 'that'. If you think about it, a collection that includes a fruit fly and Richard Dawkins and the Great Barrier Reef is an awkward set of objects to try and compare.

Douglas laughed at himself, and at his own jokes. It was one of many ingredients of his charm.

There are some oddities in the perspective with which we see the world. The fact that we live at the bottom of a deep gravity well, on the surface of a gas-covered planet going around a nuclear fireball 90 million miles away and think this to be *normal* is obviously some indication of how skewed our perspective tends to be, but we have done various things over intellectual history to slowly correct some of our misapprehensions.

This next paragraph is one of Douglas's set-pieces which will be familiar to some people here. I heard it more than once, and I thought it was more brilliant every time.

... imagine a puddle waking up one morning and thinking, 'This is an interesting world I find myself in – an interesting hole I find myself in – fits me rather neatly, doesn't it? In fact it fits me staggeringly well, must have been made to have me in it!' This is such a powerful idea that as the sun rises in the sky and the air heats up and as, gradually, the puddle gets smaller and smaller, it's still frantically hanging on to the notion that everything's going to be alright, because this world was meant to have him in it, was built to have him in it; so the moment he disappears catches him rather by surprise. I think this may be something we need to be on the watch out for.

Douglas introduced me to Lalla. They had worked together, years ago, on Dr Who, and it was she who pointed out to me that he had a wonderful childlike capacity to go straight for the wood, and never mind the trees.

If you try and take a cat apart to see how it works, the first thing you have on your hands is a non-working cat. Life is a level of complexity that almost lies outside our vision; it is so far beyond anything we have any means of understanding that we just think of it as a different class of object, a different class of matter; 'life', something that had a mysterious essence about it, was god given – and that's the only explanation we had. The bombshell comes in 1859 when Darwin publishes 'On the Origin of Species'. It takes a long time before we really get to grips with this and begin to understand it, because not only does it seem incredible and thoroughly demeaning to us, but it's yet another shock to our system to discover that not only are we not the centre of the Universe and we're not made of anything, but we started out as some kind of slime and got to where we are via being a monkey. It just doesn't read well...

I am happy to say that Douglas's acquaintance with a particular modern book on evolution, which he chanced upon in his early thirties, seems to have been something of a Damascus experience for him:

It all fell into place. It was a concept of such stunning simplicity, but it gave rise, naturally, to all of the infinite and baffling complexity of life. The awe it inspired in me made the awe that people talk about in respect of religious experience seem, frankly, silly beside it. I'd take the awe of understanding over the awe of ignorance any day.[99]

I once interviewed Douglas on television, for a programme I was making on my own love affair with science. I ended up by asking him, 'What is it about science that really gets your blood running?' And here is what he said, again impromptu, and all the more passionate for that.

The world is a thing of utter inordinate complexity and richness and strangeness that is absolutely awesome. I mean the idea that such complexity can arise not only out of such simplicity, but probably absolutely out of nothing, is the most fabulous extraordinary idea. And once you get some kind of inkling of how that might have happened – it's just wonderful. And … the opportunity to spend 70 or 80 years of your life in such a universe is time well spent as far as I am concerned.[100]

That last sentence of course has a tragic ring for us now. It has been our privilege to know a man whose capacity to make the best of a full lifespan was as great as was his charm and his humour and his sheer intelligence. If ever a man understood what a magnificent place the world is, it was Douglas. And if ever a man left it a better place for his existence, it was Douglas. It would have been nice if he'd given us the full 70 or 80 years. But by God we got our money's worth from the 49!

Eulogy for W. D. Hamilton

Delivered at the Memorial Service in
New College Chapel, Oxford, 1 July 2000

Those of us who wish we had met Charles Darwin can console ourselves: we may have met the nearest equivalent that the late
twentieth century had to offer. Yet so quiet, so absurdly modest was
he that I dare say some members of this college were somewhat
bemused to read his obituaries – and discover quite what
it was they had harboured among them all this time. The obituaries
were astonishingly unanimous. I'm going to read a sentence or two
from them, and I would add that this is not a biased sample of
obituaries. I am going to quote from 100 per cent of the obituaries
that have so far come to my notice [my emphases]:

> Bill Hamilton, who has died aged 63 after weeks in intensive care follow-
> ing a biological expedition to the Congo, was *the primary theoretical
> innovator in modern Darwinian biology, responsible for the shape of the
> subject today.* [Alan Grafen in *The Guardian*.]

> ... *the most influential evolutionary biologist of his generation.* [Matt
> Ridley in the *Telegraph*.]

> ... *one of the towering figures of modern biology* ... [Natalie Angier in the
> *New York Times*.]

> ... *one of the greatest evolutionary theorists since Darwin.* Certainly,
> where social theory based on natural selection is concerned, he was
> easily our deepest and most original thinker. [Robert Trivers in *Nature*.]

> ... *one of the foremost evolutionary theorists of the twentieth century* ...
> [David Haig, Naomi Pierce and E. O. Wilson in *Science*.]

> *A good candidate for the title of most distinguished Darwinian since
> Darwin.* [That was my offering, in *The Independent*, reprinted in *Oxford
> Today*.]

... one of the leaders of what has been called 'the second Darwinian revolution'. [John Maynard Smith in *The Times*. Maynard Smith had earlier called him, in language too informal to be repeated in *The Times* obituary, 'The only bloody *genius* we've got'.]

[Finally, Olivia Judson in *The Economist*]: All his life, Bill Hamilton played with dynamite. As a boy, he nearly died when a bomb he was building exploded too soon, removing the tips of several fingers and lodging shrapnel in his lung. As an adult, his dynamite was more judiciously placed. He blew up established notions, and erected in their stead an edifice of *ideas stranger, more original and more profound than that of any other biologist since Darwin*.

Admittedly, the largest gap in the theory left by Darwin had already been plugged by R. A. Fisher and the other 'neo-Darwinian' masters of the 1930s and 40s. But their 'Modern Synthesis' left a number of important problems unsolved – in many cases even unrecognized – and most of these were not cleared up until after 1960. It is certainly fair to say that Hamilton was the dominant thinker of this second wave of neo-Darwinism, although to describe him as a solver of problems somehow doesn't do justice to his positively creative imagination.

He frequently would bury, in throwaway lines, ideas that lesser theorists would have given their eye teeth to have originated. Bill and I were once talking termites at coffee time in the Department of Zoology. We were especially wondering what evolutionary pressure had driven the termites to become so extremely social, and Hamilton started praising 'Stephen Bartz's Theory'. 'But Bill,' I protested, 'That isn't Bartz's theory. It's your theory. You published it seven years earlier.' Gloomily, he denied it. So I ran to the library, found the relevant volume of the *Annual Review of Ecology and Systematics*, and shoved under his nose his own, buried paragraph. He read it, then conceded, in his most Eeyorish voice that, yes, it did appear to be his own theory after all. 'But Bartz expressed it better.'* As a final footnote to this story, among the

---

*This is true and I have no wish, by quoting this story, to disparage Stephen Bartz's contribution. Bill Hamilton knew, better than most, that to sketch an idea on the back of an envelope is not the same as to develop it into a full model.

people whom Bartz acknowledged in his paper, 'for helpful advice and criticism', was – W. D. Hamilton!

Similarly, Bill published his theory of the sex ratio of honeybees, not in a Note to *Nature* devoted to the topic, as a normally ambitious scientist would have done, but buried in a review of somebody else's book. This book review, by the way, carried the unmistakeably Hamiltonian title, 'Gamblers since Life Began: Barnacles, Aphids, Elms'.

The two towering achievements for which Hamilton is best known were the genetic theory of kinship, and the parasite theory of sex. But, alongside these two major obsessions, he also found time to answer, or play a major role in the cooperative answering of, a whole set of other important questions left over from the neo-Darwinian synthesis. These questions include:

Why do we grow old, and die of old age?

Why do population sex ratios sometimes depart from the normally expected 50/50? In the course of this short paper, he was one of the first to introduce the Theory of Games to evolutionary biology, a development that was of course to prove so endlessly fruitful in John Maynard Smith's hands.

Can active spite, as opposed to ordinary selfishness, be favoured by natural selection?

Why do so many animals flock, school or herd together when at risk from predators? This paper had another very characteristic title: Geometry for the Selfish Herd.

Why do animals and plants go to such lengths to disperse their progeny far and wide, even when the places they are dispersing to are inferior to the place where they already live? This work was done jointly with Robert May.

In a fundamentally selfish Darwinian world, how can cooperation evolve between unrelated individuals? This work was done jointly with the social scientist Robert Axelrod.

Why do autumn leaves turn so conspicuously red or brown? In a typically audacious – yet compelling – piece of theorizing, Hamilton suspected that the bright colour is a warning given by the tree, a warning to insects not to lay their eggs on this tree, a warning backed up by toxins just as a wasp's yellow and black stripes are backed up by a sting.

This extraordinary idea is typical of that youthful inventiveness which seemed, if anything, to increase as he grew older. It was really quite recently that he proposed a proper theory for how the hitherto rather ridiculed theory of 'Gaia' could actually be made workable in a true Darwinian model. At his burial on the edge of Wytham Wood this March, his devoted companion Luisa Bozzi spoke some beautiful words over the open grave, in which she made allusion to the astonishing central idea of this paper – that clouds are actually adaptations, made by micro-organisms for their own dispersal. She quoted Bill's remarkable article 'No stone unturned: A bug-hunter's life and death', in which he expressed a wish, when he died, to be laid out on the forest floor in the Amazon jungle and interred by burying beetles as food for their larvae.[101]

> Later, in their children, reared with care by the horned parents out of fist-sized balls moulded from my flesh, I will escape. No worm for me, or sordid fly: rearranged and multiple, I will at last buzz from the soil like bees out of a nest – indeed, buzz louder than bees, almost like a swarm of motor bikes. I shall be borne, beetle by flying beetle, out into the Brazilian wilderness beneath the stars.

Luisa read this, then added her own elegy, inspired by his cloud theory:

> Bill, now your body is lying in the Wytham woods, but from here you will reach again your beloved forests. You will live not only in a beetle, but in billions of spores of fungi and algae. Brought by the wind higher up into the troposphere, all of you will form the clouds, and wandering across the oceans, will fall down and fly up again and again, till eventually a drop of rain will join you to the water of the flooded forest of the Amazon.*

Hamilton was garlanded eventually with honours, but in a way this only underlined how slow the world was to recognize him. He won many prizes, including the Crafoord Prize and the Kyoto Prize. Yet his disturbingly candid autobiography reveals a *young* man tormented by self-doubt and loneliness. Not only did he doubt himself. He was led to doubt even whether the *questions*

---

*At the memorial service, Luisa read both these passages herself. The second passage is carved on a bench beside his grave, erected by his sister Dr Mary Bliss in his memory.

that obsessively drove him were of any interest to anybody else at all. Not surprisingly, this even occasionally led him to doubt his sanity.

The experience gave him a lifelong sympathy for underdogs, which may have motivated his recent championing of an unfashionable, not to say reviled, theory of the origin of human AIDS. As you may know, it was this that was to take him on his fateful journey to Africa this year.

Unlike other major prizewinners, Bill really needed the money. He was the despair of his financial advisers. He was interested in money only for the good that it could do, usually to others. He was hopeless at accruing the stuff, and he gave away much of what he had. It was entirely characteristic of his financial astuteness that he left a will that was generous but – unwitnessed. Equally characteristic that he bought a house in Michigan at the top of the market, and later sold it at the bottom of the market. Not only did Bill's investment fail to keep up with inflation. He actually made a substantial loss, and could not afford to buy a house in Oxford. Fortunately, the university had a nice house in its gift in Wytham village, and, with Dick Southwood, as ever, quietly taking care of him behind the scenes, Bill and his wife Christine and their family found a place where they could thrive.

Every day he cycled into Oxford from Wytham, at enormous speed. So unbecoming was this speed to his great shock of grey hair, it may have accounted for his numerous cycle accidents. Motorists didn't believe that a man of his apparent age could possibly cycle so fast, and they miscalculated, with unfortunate results. I have been unable to document the widely repeated story that on one occasion he shot into a car, landed on the back seat and said, 'Please drive me to the hospital.' But I have found reliable confirmation of the story that his startup grant from the Royal Society, a cheque for £15,000, blew out of his bicycle basket at high speed.

I first met Bill Hamilton when he visited Oxford from London in about 1969 to give a lecture to the Biomathematics Group, and I went along to get my first glimpse of my intellectual hero. I won't say it was a let-down, but he was not, to say the least, a charismatic speaker. There was a blackboard that completely covered one wall. And Bill made the most of it. By the end of the seminar, there wasn't a square inch of wall that was not smothered in equations. Since the blackboard went all the way down to the floor, he had to get on his hands and knees in order to write down there, and this made his murmuring voice even more inaudible. Finally, he stood up and surveyed his handiwork with a slight smile. After a long pause, he pointed to a particular equation (aficionados may like to know that it was the now famous 'Price Equation'[102]) and said: 'I really like that one.'

I think all his friends have their own stories to illustrate his shy and idiosyncratic charm, and these will doubtless grow into legends over time. Here's one that I have vouched for, as I was the witness myself. He appeared for lunch in New College one day, wearing a large paperclip attached to his glasses. This seemed eccentric, even for Bill, so I asked him: 'Bill, why are you wearing a paperclip on your glasses?' He looked solemnly at me. 'Do you really want to know?' he said in his most mournful tone, though I could see his mouth twitching with the effort of suppressing a smile. 'Yes,' I said enthusiastically, 'I really really want to know.' 'Well,' he said, 'I find that my glasses sit heavily on my nose when I am reading. So I use the clip to fasten them to a lock of my hair, which takes some of the weight.' Then as I laughed, he laughed too, and I can still see that wonderful smile as his face lit up with laughing at himself.

On another occasion, he came to a dinner party at our house. Most of the guests were standing around drinking before dinner, but Bill had disappeared into the next room and was investigating my bookshelves. We gradually became aware of a sort of low murmuring sound coming from the next room. 'Help.' 'Er, Help ... I think. Er, yes, Help! Help.' We finally realized that, in his own uniquely understated way, Bill was saying the equivalent of 'HEEEELLLLP!!!!!!' So we rushed in there, to find him, like Inspector Clouseau with the billiard cues, struggling desperately to balance books which were falling all around him as the shelves collapsed in his arms.

Any other scientist of his distinction would expect to be offered a first-class air fare and a generous honorarium before agreeing to go and give a lecture abroad. Bill was invited to a conference in

Russia. Characteristically, he forgot to notice that they weren't offering any air fare at all, let alone an honorarium, and he ended up not only paying for his own ticket but obliged to bribe his own way out of the country. Worse, his taxi didn't have enough petrol in its tank to get him to Moscow airport, so Bill had to help the taxi driver as he siphoned petrol out of his cousin's car. As for the conference itself, it turned out when Bill got there that there was no venue for it. Instead, the delegates went for walks in the woods. From time to time, they would reach a clearing and would stop for somebody to present a lecture. Then they'd move on and look for another clearing. Bill had the impression this was an automatic precaution to avoid bugging by the KGB. He had brought slides for his lecture, so they had to go for a *night*-time ramble, lugging a projector along. They eventually found an old barn and projected his slides on its whitewashed wall. Somehow I cannot imagine any other Crafoord Prizewinner getting himself into this situation.

His absent-mindedness was legendary, but was completely unaffected. As Olivia Judson wrote in *The Economist*, his duties at Oxford required him to give only one undergraduate lecture per year, and he usually forgot to give that. Martin Birch reports that he met Bill one day in the Department of Zoology, and apologized for forgetting to go to Bill's research seminar the day before. 'That's all right,' said Bill. 'As a matter of fact, I forgot it myself.'

I made it a habit, whenever there was a good seminar or research lecture on in the Department, to go to Bill's room five minutes before it started, to tell him about it and encourage him to go. He would look up courteously from whatever he was absorbed in, listen to what I had to say, then rise enthusiastically and accompany me to the seminar. It was no use reminding him *more* than five minutes ahead of time, or sending him written memos. He would simply become reabsorbed in whatever was his current obsession, and forget everything else. For he was an obsessive. This is surely a large contributor to his success. There were other important ingredients. I love Robert Trivers's musical analogy: 'While the rest of us speak and think in single notes, he thought in chords.' That is exactly right.

He was also a wonderful naturalist – he almost seemed to prefer the company of naturalists to that of theorists. Yet he was a much better mathematician than most biologists, and he had the mathematician's way of *visualizing* the abstract and pared-down essence of a situation before he went on to model it. Though many of his papers were mathematical, Bill was also a splendidly individual prose stylist. Here's how, in his auto-anthology, *Narrow Roads of Gene Land*,[103] he introduces the reprinting of his 1966 paper on the Moulding of Senescence by Natural Selection. He first transcribes for us a marginal note which he wrote on his own copy of his 1966 paper:

Thus ageing animal should climb *down* his evolutionary tree: young man's youthful features in trends which made *old* gorilla.

This leads his older self into a magnificently Hamiltonian set-piece:

Therefore, one last confession. I, too, am probably coward enough to give funds for 'elixir' gerontology if anyone could persuade me that there is hope: at the same time I want there to be none so that I will not be tempted. Elixirs seem to me an anti-eugenical aspiration of the worst kind and to be no way to create a world our descendants can enjoy. Thus thinking, I grimace, rub two unrequestedly bushy eyebrows with the ball of a happily still-opposable thumb, snort through nostrils that each day more resemble the horse-hair bursts of an old Edwardian sofa, and, with my knuckles not yet touching the ground, though nearly, galumph onwards to my next paper.

His poetic imagination is constantly surfacing in little asides, even in his most difficult papers. And, as you would expect, he was a great lover of poets, and carried much poetry in his head, especially A. E. Housman. Perhaps he identified his young self with the melancholy protagonist of *A Shropshire Lad*. In his review of my own first book – and can you imagine my joy at receiving a review from such a quarter? – he quoted these lines:*

From far, from eve and morning
And yon twelve-winded sky,
The stuff of life to knit me
Blew hither: here am I.

*Read at the Memorial Service by Ruth Hamilton.

Now – for a breath I tarry
Nor yet disperse apart –
Take my hand quick and tell me,
What have you in your heart.

Speak now, and I will answer;
How shall I help you, say;
Ere to the wind's twelve quarters
I take my endless way.

He ended the same review by quoting Wordsworth's well-known lines on the statue of Newton in the Antechapel of Trinity College, Cambridge. Bill didn't mean it this way, of course, but the last words of the poem fit *him* as well as they fit Newton, and I want to leave you with them.

… a mind forever
Voyaging through strange seas of thought, alone.

Snake Oil

**Foreword to the posthumous book *Snake Oil
and Other Preoccupations* by John Diamond**[104]

John Diamond gave short shrift to those among his many
admirers who praised his courage. But there are distinct kinds of
courage, and we mustn't confuse them. There's physical fortitude
in the face of truly outrageous fortune, the stoical courage to
endure pain and indignity while wrestling heroically with a
peculiarly nasty form of cancer. Diamond disclaimed this kind of
courage for himself (I think too modestly, and in any case nobody
could deny the equivalent in his wonderful wife). He even used
the subtitle *Because Cowards Get Cancer Too* for his moving and I
still think brave memoir of his own affliction.

But there's another kind of courage, and here John Diamond is
unequivocally up there with the best of them. This is intellectual
courage: the courage to stick by your intellectual principles, even
when *in extremis* and sorely tempted by the easy solace that a
betrayal might seem to offer. From Socrates through David Hume
to today, those led by reason to eschew the security blanket of
irrational superstition have always been challenged: 'It's fine for
you to talk like that now. Just wait till you are on your deathbed.
You'll soon change your tune.' The solace politely refused by
Hume (as we know from Boswell's morbidly curious deathbed
visit) was one appropriate to his time. In John Diamond's time,
and ours, it is 'alternative' miracle cures, offered when orthodox
medicine seems to be failing and may even have given up on us.

When the pathologist has read the runes; when the oracles of
X-ray, CT scan and biopsy have spoken and hope is guttering low;
when the surgeon enters the room accompanied by 'a tallish man
... looking embarrassed ... in hood and gown with a scythe over
his shoulder', it is then that the 'alternative' or 'complementary'
vultures start circling. This is their moment. This is where they

come into their own, for there's money in hope: the more desperate the hope, the richer the pickings. And, to be fair, many pushers of dishonest remedies are motivated by an honest desire to help. Their persistent importunings of the gravely ill, their intrusively urgent offers of pills and potions, have a sincerity that rises above the financial greed of the quacks they promote.

> Have you tried squid's cartilage? Establishment doctors scorn it, of course, but my aunt is still alive on squid's cartilage two years after her oncologist gave her only six months (well, yes, since you ask, she is having radiotherapy as well). Or there's this wonderful healer who practises the laying on of feet, with astonishing results. Apparently it's all a question of tuning your holistic (or is it holographic?) energies to the natural frequencies of organic (or is it orgonic?) cosmic vibrations. You've nothing to lose, you might as well try it. It's £500 for a course of treatment, which may sound a lot but what's money when your life is at stake?

As a public figure who wrote, movingly and personally, about the horrible progress of his cancer, John Diamond was more than usually exposed to such siren songs: actively inundated with well-intentioned advice and offers of miracles. He examined the claims, looked for evidence in their favour, found none, saw further that the false hopes they aroused could actually be damaging – and he retained this honesty and clarity of vision to the end. When my time comes, I do not expect to show a quarter of John Diamond's physical fortitude, disavow it though he might. But I very much hope to use him as my model when it comes to intellectual courage.

The obvious and immediate countercharge is one of arrogance. Far from being rational, wasn't John Diamond's 'intellectual courage' really an unreasoning overconfidence in science, a blind and bigoted refusal to contemplate alternative views of the world, and of human health? No, no and no. The accusation would stick if he had bet on orthodox medicine simply because it is orthodox, and shunned alternative medicine simply because it is alternative. But of course he did no such thing. For his purposes (and mine), scientific medicine is *defined* as the set of practices which submit themselves to the ordeal of being *tested*. Alternative medicine is defined as that set of practices which cannot be tested, refuse to be tested or consistently fail tests. If a healing technique is demonstrated to have curative properties in properly controlled double-blind trials, it ceases to be alternative. It simply, as Diamond explains, becomes medicine. Conversely, if a technique devised by the President of the Royal College of Physicians consistently fails in double-blind trials, it will cease to be a part of 'orthodox' medicine. Whether it will then become 'alternative' will depend upon whether it is adopted by a sufficiently ambitious quack (there are always sufficiently gullible patients).

But isn't it still an arrogance to demand that our method of *testing* should be the scientific method? By all means use scientific tests for scientific medicine, it may be said. But isn't it only fair that 'alternative' medicine should be tested by 'alternative' tests? No. There is no such thing as an alternative test. Here Diamond takes his stand, and he is right to do so.

Either it is true that a medicine works or it isn't. It cannot be false in the ordinary sense but true in some 'alternative' sense. If a therapy or treatment is anything more than a placebo, properly conducted double-blind trials, statistically analysed, will eventually bring it through with flying colours. Many candidates for recognition as 'orthodox' medicines fail the test and are summarily dropped. The 'alternative' label should not (though, alas, it does) provide immunity from the same fate.

Prince Charles has recently called for ten million pounds of government money to be spent researching the claims of 'alternative' or 'complementary' medicine. An admirable suggestion, although it is not immediately clear why government, which has to juggle competing priorities, is the appropriate source of money, given that the leading 'alternative' techniques have already been tested – and have failed – again and again and again. John Diamond tells us that the alternative medicine business in Britain has a turnover measured in billions of pounds. Perhaps some small fraction of the profits generated by these medicines could be diverted into testing whether they actually work. This, after all, is what 'orthodox' pharmaceutical companies are expected to do. Could it be that purveyors of alternative medicine know all too well what the upshot of properly conducted trials would be? If so, their reluctance to fund their own nemesis is all too understandable. Nevertheless, I hope this research money will come from somewhere, perhaps from Prince Charles's own charitable resources, and I would be happy to serve on an advisory committee to disburse it, if invited to do so. Actually, I suspect that ten million pounds' worth of research is more than would be necessary to see off most of the more popular and lucrative 'alternative' practices.

How might the money be spent? Let's take homeopathy as an example, and let us suppose that we have a large enough fraction of the grant to plan the experiment on a moderately large scale. Having given their consent, 1000 patients will be separated into 500 experimentals (who will receive the homeopathic dose) and 500 controls (who will not). Bending over backwards to respect the 'holistic' principle that every individual must be treated as an individual, we shall not insist on giving all experimental subjects the same dose. Nothing so crude. Instead, every patient in the trial shall be examined by a certified homeopath, and an individually tailored therapy prescribed. The different patients need not even receive the same homeopathic substance.

But now comes the all-important double-blind randomization. After every patient's prescription has been written, half of the patients, at random, will be designated controls. The controls will not in fact receive their prescribed dose. Instead, they will be given a dose which is identical in all respects to the prescribed dose but with one crucial difference. The supposed active ingredient is omitted from its preparation. The randomizing will be done by computer, in such a way that nobody will know which patients are experimentals and which controls. The patients themselves won't know; the therapists won't know; the pharmacists preparing the doses won't know, and the doctors judging the results won't know. The bottles of medicine will be identified only by impenetrable code numbers. This is vitally important because nobody denies placebo effects: patients who think they are getting an effective cure feel better than patients who think the opposite.

Each patient will be examined by a team of doctors and homeopaths, both before and after the treatment. The team will write down their judgement for each patient: has this patient got better, stayed the same, or got worse? Only when these verdicts have all been written down and sealed will the randomizing codes in the computer be broken. Only then will we know which patients had received the homeopathic dose and which the control placebo. The results will be analysed statistically to see whether the homeopathic doses had any effect one way or the other. I know which result my shirt is on, but – this is the beauty of good science – I cannot bias the outcome. Nor can the homeopaths who are betting on the opposite. The double-blind experimental design disempowers all such biases. The experiment can be performed by advocates or sceptics, or both working together, and it won't change the result.

There are all sorts of details by which this experimental design could be made more sensitive. The patients could be sorted into 'matched pairs', matched for age, weight, sex, diagnosis, prognosis and preferred homeopathic prescription. The only consistent difference is that one member of each pair is randomly and secretly designated a control, and given a placebo. The statistics then specifically compare each experimental individual with his matched control.

The ultimate matched-pairs design is to use each patient as his own control, receiving the experimental and the control dose successively, and never knowing when the change occurs. The order of administering the two treatments to a given patient would be determined at random, a different random schedule for different patients.

'Matched pairs' and 'own control' experimental designs have the advantage of increasing the sensitivity of the test. Increasing, in other words, the chance of yielding a statistically significant success for homeopathy. Notice that a statistically significant success is not a very demanding criterion. It is not necessary that every patient should feel better on the homeopathic dose than on the control. All we are looking for is a slight advantage to homeopathy over the blind control, an advantage which, however slight, is too great to be attributable to luck, according to the standard methods of statistics. This is what is routinely demanded of orthodox medicines before they are allowed to be advertised and sold as curative. It is rather less than is demanded by a prudent pharmaceutical company before it will invest a lot of money in mass production.

Now we come to an awkward fact about homeopathy in particular, dealt with by John Diamond, but worth stressing here. It is a fundamental tenet of homeopathic theory that the active ingredient – arnica, bee venom, or whatever it is – must be successively diluted some large number of times, until – all calculations agree – there is not a single molecule of that ingredient remaining. Indeed, homeopaths make the daringly paradoxical claim that the more dilute the solution the more potent its action. The investigative conjuror James Randi has calculated that, after a typical sequence of homeopathic 'successive' dilutions, there would be one molecule of active ingredient in a vat the size of the solar system! (Actually, in practice, there will be more stray molecules knocking around even in water of the highest attainable purity.)

Now, think what this does. The whole rationale of the experiment is to compare experimental doses (which include the 'active' ingredient) with control doses (which include all the same ingredients except the active one). The two doses must look the same, taste the same, feel the same in the mouth. The only respect in which they differ must be the presence or absence of the putatively curative ingredient. But in the case of homeopathic medicine, the dilution is such that there is no difference between the experimental dose and the control! Both contain the same number of molecules of the active ingredient – zero, or whatever is the minimum attainable in practice. This seems to suggest that a double-blind trial of homeopathy cannot, in principle, succeed. You could even say that a successful result would be diagnostic of a failure to dilute sufficiently!

There is a conceivable loophole, much slithered through by homeopaths ever since this embarrassing difficulty was brought to their attention. The mode of action of their remedies, they say, is not chemical but physical. They agree that not a single molecule of the active ingredient remains in the bottle that you buy, but this only matters if you insist on thinking chemically. They believe that, by some physical mechanism unknown to physicists, a kind of 'trace' or 'memory' of the active molecules is imprinted on the water molecules used to dilute them. It is the physically imprinted template on the water that cures the patient, not the chemical nature of the original ingredient.

This is a scientific hypothesis in the sense that it is testable. Easy to test, indeed, and although I wouldn't bother to test it myself, this is only because I think our finite supply of time and money would be better spent testing something more plausible. But any homeopath who really believes his theory should be beavering away from dawn to dusk. After all, if the double-blind trials of patient treatments came out reliably and repeatably positive, he would win a Nobel Prize not only in Medicine but in Physics as well. He would have discovered a brand-new principle of physics, perhaps a new fundamental force in the universe. With such a prospect in view, homeopaths must surely be falling over each other in their eagerness to be first into the lab, racing like alternative Watsons and Cricks to claim this glittering scientific crown. Er, actually no they aren't. Can it be that they don't really believe their theory after all?

At this point we scrape the barrel of excuses. 'Some things are true on a human level, but they don't lend themselves to scientific testing. The sceptical atmosphere of the science lab is not conducive to the sensitive forces involved.' Such excuses are commonly trotted out by practitioners of alternative therapies, including those that don't have homeopathy's peculiar difficulties of principle but which nevertheless consistently fail to pass double-blind tests in practice. John Diamond is a pungently witty writer, and one of the funniest passages of this book is his description of an experimental test of 'kinesiology' by Ray Hyman, my colleague on CSICOP (the Committee for the Scientific Investigation of Claims of the Paranormal).

As it happens, I have personal experience of kinesiology. It was used by the one quack practitioner I have – to my shame – consulted. I had ricked my neck. A therapist specializing in manipulation had been strongly recommended. Manipulation can undoubtedly be very effective, and this woman was available at the weekend, when I didn't like to trouble my normal doctor. Pain and an open mind drove me to give her a try. Before she began the manipulation itself, her diagnostic technique was kinesiology. I had to lie down and stretch out my arm, and she pushed against it, testing my strength. The key to the diagnosis was the effect of vitamin C on my arm-wrestling performance. But I wasn't asked to imbibe the vitamin. Instead (I am not exaggerating, this is the literal truth), a sealed bottle containing vitamin C was placed on my chest. This appeared to cause an immediate and dramatic increase in the strength of my arm, pushing against hers. When I expressed my natural scepticism, she said happily, 'Yes, C is a marvellous vitamin, isn't it!' Human politeness stopped me walking out there and then, and I even (to avoid hassle) ended up paying her lousy fee.

What was needed (I doubt if that woman would even have understood the point) was a series of double-blind trials, in which neither she nor I was allowed to know whether the bottle contained the alleged active ingredient or something else. This was what Professor Hyman, in John Diamond's hilarious description of a similar case, undertook. When, predictably, the 'alternative' technique ignominiously flunked the double-blind test, its practitioner delivered himself of the following immortal response: 'You see? That is why we never do double-blind testing any more. It never works!'

A large part of the history of science, especially medical science, has been a progressive weaning away from the superficial seductiveness of individual stories that seem – but only seem – to show a pattern. The human mind is a wanton storyteller and, even more, a profligate seeker after pattern. We see faces in clouds and tortillas, fortunes in tea leaves and planetary movements. It is quite difficult to prove a real pattern as distinct from a superficial illusion. The human mind has to learn to mistrust its native tendency to run away with itself and see pattern where there is only randomness. That is what statistics are for, and that is why no drug or therapeutic technique should be adopted until it has been proved by a statistically analysed experiment, in which the fallible pattern-seeking proclivities of the human mind have been systematically taken out of the picture. Personal stories are never good evidence for any general trend.

In spite of this, doctors have been heard to begin a judgement with something like, 'The trials all say otherwise, but in *my* clinical experience...' This might constitute better grounds for changing your doctor than a suable malpractice! That, at least, would seem to follow from all that I have been saying. But it is an exaggeration. Certainly, before a medicine is certified for wide use, it must be properly tested and given the imprimatur of statistical significance. But a mature doctor's clinical experience is at least an excellent guide to which hypotheses might be worth going to the trouble and expense of testing. And there's more that can be said. Rightly or wrongly (often rightly) we actually do take the personal judgement of a respected human individual seriously. This is so with aesthetic judgements, which is why a famous critic can make or break a play on Broadway or Shaftesbury Avenue. Whether we like it or not, people are swayed by anecdote, by the particular, by the personal.

And this, almost paradoxically, helps to make John Diamond such a powerful advocate. He is a man whom we like and admire for his personal story, and whose opinions we want to read because he expressed them so well. People who might not listen to a set of nameless statistics, intoned by a faceless scientist or doctor, will listen to John Diamond, not just because he writes engagingly, but because he was dying while he wrote and he knew it: dying in spite of the best efforts of the very medical practices he was defending against opponents whose only weapon is anecdote. But there is really no paradox. He may gain our ear because of his singular qualities and his human story. But what we hear when we listen to him is not anecdotal. It stands up to rigorous examination. It would be sensible and compelling in its own right even if its author had not previously earned our admiration and our affection.

John Diamond was never going to go gently into that good night. When he did go it was with guns blazing, for the splendidly polemical chapters of *Snake Oil* occupied him right up to the end, working against ... not so much the clock as time's wingèd chariot itself. He does not rage against the dying of the light, nor against his wicked cancer, nor against cruel fate. What would be the point, for what would they care? His targets are capable of wincing when hit. They are targets that deserve to be hit hard, targets whose neutralization would leave the world a better place: cynical charlatans (or honestly foolish dreamers) who prey on gullible unfortunates. And the best part is that although this gallant man is dead, his guns are not silenced. He left a strong emplacement. This posthumous book launches his broadside. Open fire, and don't stop.

5

EVEN THE RANKS OF TUSCANY

Stephen Jay Gould and I did not tire the sun with talking and send him down the sky. We were cordial enough when we met, but it would be disingenuous to suggest that we were close. Our academic differences have even been spun out to book length, by the philosopher Kim Sterelny in *Dawkins vs Gould: Survival of the Fittest*,[105] while Andrew Brown, in *The Darwin Wars: How Stupid Genes Became Selfish Gods*,[106] goes so far as to divide modern Darwinians into 'Gouldians' and 'Dawkinsians'. Yet, despite our differences, it is not just the respect due to the dead that leads me to include in this book a section on Stephen Gould with a largely positive tone.

'And even the ranks of Tuscany' (Steve would have completed the quotation from his formidable literary memory) 'Could scarce forebear to cheer'. Macaulay[107] celebrated the admiration that can unite enemies in death. Enemies is too strong a word for a purely academic dispute, but admiration is not, and we were shoulder to shoulder on so much. In his review of my own *Climbing Mount Improbable*, Steve invoked a collegiality between us, which I reciprocated, in the face of a shared enemy:[108]

> In this important uphill battle for informing a hesitant (if not outrightly hostile) public about the claims of Darwinian evolution, and for explaining both the beauty and power of this revolutionary view of life, I feel collegially entwined with Richard Dawkins in a common enterprise.

He was never ashamed of his immodesty, and I hope I may be forgiven for sharing with my readers the one occasion when he was good enough to include me in it: 'Richard and I are the two people who write about evolution best ...'[109] There was a 'but' of course, but I must press on.

The book reviews that follow, spaced many years apart, show what

I hope will be read as an equal collegiality, even where they are critical. *Ever Since Darwin* was the first collection of Gould's celebrated essays from *Natural History*. It set the tone for all ten of them, and the 'barbed rave' tone of **Rejoicing in Multifarious Nature** (5.1) could also serve for any.

The Art of the Developable (5.2), though written in 1983, has not previously been published. It is a joint review of Peter Medawar's *Pluto's Republic* and the third of Gould's collections of essays from *Natural History*. It was commissioned by the *New York Review of Books* but eventually, for reasons that I no longer recall, the publication fell through. Years later I sent the review to Steve, and he expressed warm disappointment that it had never been published. Medawar was one of my intellectual heroes, and Gould's too: it was another thing we had in common. My title, 'The Art of the Developable', unites Medawar's *Art of the Soluble*[110] with Gould's long interest in the evolution of development.

Wonderful Life is, in my view, a beautiful and a misguided book. It is also misguiding: its enthusiastic rhetoric leading other authors to absurd conclusions far beyond Dr Gould's intentions. I developed this aspect fully in 'Huge Cloudy Symbols of a High Romance', one of the chapters of my *Unweaving the Rainbow*. Reprinted here as **Hallucigenia, Wiwaxia and Friends** (5.3), the title given it by the *Sunday Telegraph*, is my review of *Wonderful Life* itself.

Human Chauvinism and Evolutionary Progress (5.4) is my review of *Full House*, a book that was renamed by the British publishers *Life's Grandeur*. The review was published as a matched pair with Steve's review of *Climbing Mount Improbable*. The Editor of *Evolution* thought it would be amusing to invite each of us to review the other's book simultaneously, knowing the existence, but not the content of the other's review. Gould's review had the characteristic title 'Self-help for a hedgehog stuck on a molehill'. *Full House* is all about the idea of progress in evolution. I agree with Gould's objections to progress as he saw it. But in this review I develop two alternative meanings of progress which I think are important and are not vulnerable to his objections. My intention was not just to review a book but to make a contribution to evolutionary thinking.

Stephen Gould was my exact contemporary but I always thought of him as senior, probably because his prodigious learning seemed to belong to a more cultivated era. His lifelong colleague Niles

Eldredge, who was kind enough to send me the text of his moving eulogy, said that he had lost an elder brother. Years ago it seemed natural to me to ask Steve's advice when I was travelling in America and was invited to have a televised 'debate' with a creationist. He said that he always refused such invitations, not because he was afraid of 'losing' the debate (the idea is laughable) but for a subtler reason which I accepted and never forgot. Shortly before his last illness began, I wrote to him, reminding him of his advice to me and proposing that we might publish a joint letter, offering the same advice to others. He enthusiastically agreed, and suggested that I should prepare a draft on which we could later work together. I did so but, sadly, 'later' never came. When I heard of his sudden death, I wrote to Niles Eldredge, asking if he thought Steve would have wished me to publish the letter anyway. Niles encouraged me to do so, and, as **Unfinished Correspondence with a Darwinian Heavyweight** (5.5), it closes this section.

For good or ill, Steve Gould had a huge influence on American scientific culture, and on balance the good came out on top. It is pleasing that, just before his death, he managed to complete both his *magnum opus* on evolution and his ten-volume cycle of essays from *Natural History*. Although we disagreed about much, we shared much too, including a spellbound delight in the wonders of the natural world, and a passionate conviction that such wonders deserve nothing less than a purely naturalistic explanation.

5.1

Rejoicing in Multifarious Nature[111]

Review of *Ever Since Darwin* by S. J. Gould

'The author shows us what is revealed when we remove the blinkers which Darwin stripped from biology a century ago.' Some overkill there, or an excitingly paradoxical striptease technique? The first essay in the book discusses Darwin's own coyness in not revealing his theory until 20 years after he thought of it, and I shall return to this. The quotation from the jacket blurb gives a false impression, for Stephen Gould's writing is elegant, erudite, witty, coherent and forceful. He is also, in my opinion, largely right. If there are elements of paradox and overkill in Dr Gould's intellectual position, they are not to be found within these covers. *Ever Since Darwin* is a collection of essays which first appeared as a regular monthly column in *Natural History*. Skilfully edited to flow in eight main sections, the 33 essays, of which I can mention only a sample, reinforce my feeling that scientific journalism is too important to be left to journalists, and encourage my hope that true scientists may be better at it than journalists anyway.

Gould's collection begins to bear comparison with P. B. Medawar's immortal *The Art of the Soluble*. And if his style does not quite make the reader chortle with delight and rush out to show somebody – anybody – the way Medawar's does, Gould is to be thanked for some memorable lines. No doubt puritan killjoys of Science for 'The People' will denounce the vivid and helpful anthropomorphism in 'Reproduce like hell while you have the ephemeral resource, for it will not last long and some of your progeny must survive to find the next one.' But on second thoughts they may be too busy plotting the abolition of slavery in ants, or brooding over the deviationism of:

Natural selection dictates that organisms act in their own self-interest ...
They 'struggle' continuously to increase the representation of their genes
at the expense of their fellows. And that, for all its baldness, is all there
is to it; we have discovered no higher principle in nature.

Ever since Darwin we have known why we exist and we have
known at least how to set about explaining human nature. I agree
that natural selection is 'the most revolutionary notion in the
history of biology' and I would toy with substituting 'science' for
'biology'. Childishly simple as it is, nobody thought of it until
centuries after far more complicated ideas had become common
currency, and it is still the subject of misunderstanding and even
apathy among educated people. A microcosm of this historical
enigma is the subject of Gould's first essay. Just as humankind
waited centuries longer than our hindsight deems necessary
before discovering natural selection, so Darwin delayed his own
publication 20 years after he first thought of the theory in 1838.
Gould's explanation is that Darwin was afraid of the psychological implications of his idea. He saw what Wallace would never
admit, that the human mind itself must be a material product of
natural selection. Darwin, in fact, was a scientific materialist.

In another essay Gould is encouraged by the genetic closeness
of humans and chimpanzees to speculate that 'inter-breeding
may well be possible'. I doubt it, but it is a pleasing thought and
Gould surely exaggerates when he rates it 'the most ... ethically
unacceptable scientific experiment I can imagine'. For my ethics,
far less acceptable experiments are conceivable, and actually done
in animal physiology laboratories every day, and a chimp/human
hybridization would provide exactly the come-uppance that
'human dignity' needs. Gould is, in general, rather good at
puncturing human speciesist vanity, and in particular he will
have nothing to do with the myth that evolution represents
progress towards man. This scepticism informs his valuable
account of 'Bushes and ladders in human evolution', and fires his
scorn for attempts to rank human races as primitive or advanced.

He returns to the attack on progress in the very different guise
of the theory of orthogenesis, the idea that evolutionary trends
have their own internal momentum which eventually drives

lineages extinct. His telling of the classic Irish Elk story gains freshness from his intimacy with the fossils of the Dublin Museum and gives the lie to the myth that palaeontology is dry and dull. His conclusion that the proverbially top-heavy antlers were important in social life is surely right, but he may under-estimate the role of within-species competition in driving species extinct. Large antlers could directly have caused the extinction of the Irish Elk while at the same time, right up to the moment of extinction, individuals with relatively large antlers were out-reproducing individuals with relatively small antlers. I would like to see Gould come to terms with the 'orthoselective' impact of 'arms races' both between and within species. He seems to approach this in his essays on the 'Cambrian explosion'.

Natural history can be sold for its intrinsic fascination, but it is much better used to make a point. Gould tells us about a fly that eats its mother from the inside, about 17-year cicadas and 120-year bamboos, and about uncanny fish-decoying mussels. He employs the useful trick of first opening the reader's mind by boggling it, then filling it with the important biological principle. One principle I would have liked to have heard more on is that of the limitation of evolutionary perfection: 'Orchids are Rube Goldberg machines; a perfect engineer would certainly have come up with something better' (Rube Goldberg is the American Heath Robinson). My own favourite example, inherited from an under-graduate tutor, is the recurrent laryngeal nerve. It starts in the head, goes down into the chest, loops round the aorta, then goes straight back into the head again. In a giraffe this detour must be wasteful indeed. The human engineer who first designed the jet engine simply threw the old propeller engine out and started afresh. Imagine the contraption he would have produced if he had been constrained to 'evolve' his jet engine by changing a propeller engine one bit at a time, nut by nut and bolt by bolt!

While on the problem of perfection, I think Gould exaggerates the relevance of 'neutral mutations'. Molecular geneticists are understandably interested in DNA changes as molecular events, and any that have no effect on protein function may reasonably be called neutral mutations. But to a student of whole organism biology they are less than neutral; they are not, in any interesting sense, mutations at all! If the molecular neutralists are right, their kind of neutral mutation will forever be hidden from the field biologist and from natural selection. And if a field biologist actually sees variation in phenotypes, the question of whether that variation could be selectively neutral cannot be settled in the biochemistry laboratory.

Several essays touch on aspects of the relationship of Darwinism to human society and politics. There is much humane good sense here and I agree with most of it. Although 'sociobiology' is inspiring excellent research, Gould is right that it has also led to some second-rate bandwagoneering. 'But was there ever a dog that praised his fleas?' asked Yeats. Perhaps a dog may be held responsible for the fleas he sheds, but only to a small extent. At the AAAS meeting in Washington in 1978, Gould and I witnessed an organized attack on his most distinguished Harvard colleague.* Gould well deserved his ovation for the apt Lenin quote with which he disowned the rabble. But as he watched those pathetic fleas ineffectually hopping around the stage chanting, of all things, 'genocide', did he wonder with a little itch of conscience on which dog they had been sucking?

The Epilogue is forward-looking and whets our appetite for Volume 2, which I earnestly hope will be forthcoming.† One theme which I know Gould has already carried further in his *Natural History* column is his dislike for 'the ultimate atomism' of regarding organisms as 'temporary receptacles ... no more than instruments that genes use to make more genes like themselves'.[112] In describing this as 'metaphorical nonsense' Gould underestimates the sophistication of the idea, first cogently expressed in its modern form by George C. Williams.[113] The dispute is largely semantic. Inclusive fitness is defined in such a way that to say 'the individual works so as to maximize its inclusive fitness' is equivalent to saying 'the genes work so as to maximize their survival'. The two forms are each valuable for different purposes. Both contain an element of personification; it is dangerously

---

*A glassful of water was thrown sideways at Professor E. O. Wilson (subsequently exaggerated in various accounts to a 'pitcher of iced water, poured over him').
†In fact, ten volumes were eventually published, the last one, *I have Landed*, at the time of his death.

easier to personify organisms than to personify genes. The gene selection idea is not naively atomistic, as it recognizes that genes are selected for their capacity to interact productively with the other genes with which they are most likely to share 'receptacles'; this means the other genes of the gene pool; and the gene pool may therefore come to resemble a 'homeostatically buffered system' tending to return to (one of) its evolutionarily stable state(s). Irrevocable determination by genes is no part of the idea nor is anything remotely approaching a 'one gene, one trait' mapping from genotype to phenotype. In any case it has nothing to do with 'supreme confidence in universal adaptation', which is as likely to be found among devotees of 'individual selection' or 'species selection'.

'I will rejoice in the multifariousness of nature and leave the chimera of certainty to politicians and preachers': a resounding conclusion to a stimulating book – the work of a free and imaginative scientific mind. The final, sad paradox is this. How can a mind capable of such rejoicing, open enough to contemplate the shifting splendour of three thousand million years, moved by the ancient poetry written in the rocks, how can such a mind not be bored by the drivelling ephemera of juvenile pamphleteers and the cold preaching of spiteful old hardliners? No doubt they are right that science is not politically neutral. But if, to them, that is the most important thing about science, just think what they are missing! Stephen Gould is well qualified, and strategically placed, to strip away even those dark blinkers and dazzle those poor unpractised eyes.

5.2

The Art of the Developable

**Review of *Pluto's Republic* by Peter Medawar[114] and
Hen's Teeth and Horse's Toes by Stephen Jay Gould[115]**

The acknowledged master of biological belles lettres has long been Sir Peter Medawar. If there is a younger biologist or an American biologist that bears comparison, it is probably in both cases Stephen Jay Gould. It was therefore with anticipation that I received these two collections of essays, reflections by leading and highly literate biologists on their subject and its history and philosophy.

Pluto's Republic is one of those titles that cannot be mentioned without an immediate explanation, and Sir Peter begins thus:

> A good many years ago a neighbour whose sex chivalry forbids me to disclose [it takes a Medawar to get away with this kind of thing nowadays] exclaimed upon learning of my interest in philosophy: 'Don't you just adore Pluto's Republic?' Pluto's Republic has remained in my mind ever since as a superlatively apt description of that intellectual underworld which so many of the essays in this volume explore. We each populate Pluto's Republic according to our own prejudices ...

Here I nursed a mischievous half-hope that Stephen Gould might be found among the denizens of Medawar's private underworld – his more sanctimonious cosignatories of a notorious letter to the *New York Review of Books* about 'sociobiology' (13 November 1975) are prominent in mine. But Gould is several cuts above those former associates of his and he is not among Medawar's targets. Indeed, they share many targets, IQ-metricians for instance.

Most of the essays in *Pluto's Republic* have appeared twice before, first as book reviews or transcripts of lectures, then in previous anthologies such as *The Art of the Soluble* and *The Hope of Progress*,[116] which were presumably reviewed at the time. Although I shall therefore give *Pluto's Republic* less than half my space in this joint review, I vigorously repudiate any mutterings about such second order anthologizing being too much of a good thing. The earlier books have long been out of print, and I have been scouring the second-hand bookshops ever since my own *The Art of the Soluble* was stolen. I discovered when I reread them here that I had many favourite passages word-perfect in memory. Who indeed could forget the opening sentence of the 1968 Romanes Lecture, 'Science and Literature'? 'I hope I shall not be thought ungracious if I say at the outset that nothing on earth would have induced me to attend the kind of lecture you may think I am about to give.' At the time this prompted the apt rejoinder from John Holloway: 'This lecturer can never have been thought ungracious in his life.'

Or listen to Medawar on another great biologist, Sir D'Arcy Thompson:

> ... He was a famous conversationalist and lecturer (the two are often thought to go together, but seldom do), and the author of a work which, considered as literature, is the equal of anything of Pater's or Logan Pearsall Smith's in its complete mastery of the bel canto style. Add to all this that he was over six feet tall, with the build and carriage of a Viking and with the pride of bearing that comes from good looks known to be possessed.

The reader may be hazy about Logan Pearsall Smith and Pater, but he is left with the overwhelming impression (since he probably is familiar with the idiom of P. G. Wodehouse) of a style that is undoubtedly bel, and may very well be canto. And there is more of Medawar in the passage quoted than Medawar himself realized.

Medawar continually flatters his readers, implying in them an erudition beyond them, but doing it so that they almost come to believe in it themselves:

> 'Mill,' said John Venn in 1907, has 'dominated the thought and study of intelligent students to an extent which many will find it hard to realise at the present day'; yet he could still take a general familiarity with Mill's views for granted ...

The reader scarcely notices that Medawar himself is still taking a general familiarity with Mill's views for granted, although in the reader's own case it may be far from justified. 'Even George Henry Lewes found himself unable to propound his fairly sensible views on hypotheses without prevarication and pursing of the lips.' The reader's knowing chuckle is out before he realizes that actually he is in no position to respond knowingly to that 'even'.

Medawar has become a sort of chief spokesman for 'The Scientist' in the modern world. He takes a less doleful view of the human predicament than is fashionable, believing that hands are for solving problems rather than for wringing. He regards the scientific method – in the right hands – as our most powerful tool for 'finding out what is wrong with [the world] and then taking steps to put it right'. As for the scientific method itself, Medawar has a good deal to tell us, and he is well qualified to do so. Not that being a Nobel Prize-winner and a close associate of Karl Popper is in itself an indication that one will talk sense: far from it when you think of others in that category. But Medawar not only is a Nobel Prize-winner, he *seems* like a Nobel Prize-winner; he is everything we think a Nobel Prize-winner ought to be. If you have never understood why scientists like Popper, try Medawar's exposition of the philosophy of his 'personal guru'.

He read Zoology at Oxford, and early in his career made important contributions to classical Zoology, but was soon drawn into the highly populated and highly financed world of medical research. Inevitably, his associates have been molecular and cell biologists, but he seldom had any truck with the molecular chauvinism which plagued biology for two decades. Medawar has a good appreciation of biology at all levels.

He has also inevitably associated with doctors, and the pre-occupations and sympathies of a doctor pervade several of these essays, for example his sensitive reviews of books on cancer and psychosomatic heart disease. I especially enjoyed his blistering contempt for psychoanalysis: not a lofty, detached contempt for any ordinary pretentious drivel, but a committed contempt, fired by a doctor's concern. Psychoanalysts have even had their say over the puzzle of Darwin's long illness, and Medawar is at his withering best in telling us about it.

For Good, there is a wealth of evidence that unmistakably points to the idea that Darwin's illness was 'a distorted expression of the aggression, hate and resentment felt, at an unconscious level, by Darwin towards his tyrannical father'. These deep and terrible feelings found outward expression in Darwin's touching reverence towards his father and his father's memory, and in his describing his father as the kindest and wisest man he ever knew: clear evidence, if evidence were needed, of how deeply his true inner sentiments had been repressed.

Medawar, when he smells pretentious pseudoscience, is a dangerous man. His famous annihilation of Teilhard de Chardin's *The Phenomenon of Man* might have been thought an unfair attack on the dead, but for the extraordinary influence Teilhard exerted (and still exerts: Stephen Gould tells us that two journals established to discuss his ideas still flourish) over legions of the gullible including, I am afraid, my juvenile self. I would love to quote huge chunks of what is surely one of the great destructive book reviews of all time, but must content myself with two sentences from Medawar's typically barbed explanation of the popular appeal of Teilhard.

> Just as compulsory primary education created a market catered for by cheap dailies and weeklies, so the spread of secondary and latterly of tertiary education has created a large population of people, often with well-developed literary and scholarly tastes who have been educated far beyond their capacity to undertake analytical thought ... [*The Phenomenon of Man*] is written in an all but totally unintelligible style, and this is construed as prima-facie evidence of profundity.

Medawar's Herbert Spencer Lecture, and his review of Arthur Koestler's *Act of Creation*, are more respectful of his victims, but pretty punchy nevertheless. His review of Ronald Clark's *Life of J. B. S. Haldane* is enlivened by personal reminiscence, and reveals a sort of affection for the old brute which seems to have been reciprocated.

> I remember Haldane's once going back on a firm promise to chair a lecture given by a distinguished American scientist on the grounds that it would be too embarrassing for the lecturer: he had once been the victim of a sexual assault by the lecturer's wife. The accusation was utterly ridiculous and Haldane did not in the least resent my saying so. He didn't want to be bothered with the chairmanship, and could not bring himself to say so in the usual way.

But if Haldane did not in the least resent Medawar's saying so, one cannot help wondering whether this was only because Medawar must have been one of the very few people Haldane ever met who could look him levelly in the eye, on equal terms intellectually. Peter Medawar is a giant among scientists and a wicked genius with English prose. Even if it annoys you, you will not regret reading *Pluto's Republic*.

In 1978 the Reviews Editor of a famous scientific journal, whose nature prudence forbids me to disclose, invited me to review Stephen Jay Gould's *Ever Since Darwin*, remarking that I could 'get my own back' on opponents of 'genetic determinism'. I don't know which annoyed me more: the suggestion that I favoured genetic 'determinism' (it is one of those words like sin and reductionism: if you use it at all you are against it) or the suggestion that I might review a book for motives of revenge. The story warns my readers that Dr Gould and I are supposed to be on opposite sides of some fence or other. In the event, I accepted the commission and gave the book what could fairly be described as a rave review, even, I think, going so far as to praise Gould's style as a creditable second best to Peter Medawar's.*

I feel inclined to do the same for *Hen's Teeth and Horse's Toes*. It is another collection of essays reprinted from Gould's column in *Natural History*. When you have to turn these pieces out once a month you must pick up some of the habits of the professional working to a deadline – this is not a criticism, Mozart did the same. Gould's writing has something of the predictability that we enjoy in Mozart, or in a good meal. His volumes of collected essays, of which this is the third, are put together to a recipe: one part biological history, one part biological politics (less if we are lucky), and one part (more if we are lucky) vignettes of biological wonder, the modern equivalent of a mediaeval bestiary but with interesting scientific morals instead of boring pious ones. The essays themselves, too, often seem to follow a formula or menu. As appetizer there is the quotation from light opera or the classics, or sometimes its place is taken by a piece of reassuring nostalgia; a reminiscence from a normal, happy, very American childhood world of baseball stars and Hershey bars and Bar Mitzvahs – Gould, we learn, is not just one of your pointy-headed

*See 'Rejoicing in Multifarious Nature' (pp. 222–26).

intellectuals but a regular guy. This homely informality softens the conspicuous erudition of the main course – the fluency in several languages, the almost Medawarian familiarity with literature and humanities – and even gives it a certain (un-Medawarian) charm (compare Gould himself on Louis Agassiz: '... the erudition that has so charmed American rustics ...').

Gould's own respect for Medawar is evident. The idea of science as 'the art of the soluble' provides the punchline for at least four of the essays: 'We may wallow forever in the thinkable; science traffics in the doable', '... science deals in the workable and soluble'; and two essays end with explicit quotations of the phrase. His view of Teilhard de Chardin's style is similar to Medawar's: '... difficult, convoluted writing may simply be fuzzy, not deep'. If he gives Teilhard's philosophy a slightly more sympathetic hearing, he is probably just making amends for his delightfully mischievous thesis that the young Teilhard connived in the Piltdown hoax. For Medawar, Teilhard's accepted role as one of the principal victims of the joke is just more evidence that he was

> in no serious sense a thinker. He has about him that innocence which makes it easy to understand why the forger of the Piltdown skull should have chosen Teilhard to be the discoverer of its canine tooth.

Gould's case for the prosecution is a fascinating piece of detective work which I will not spoil by attempting to summarize it. My own verdict is a Scottish 'non-proven'.

In whatever underworld the Piltdown forger languishes, he has a lot to answer for. Only last month an acquaintance, whose sex the grammar of English pronouns will probably force me to disclose, exclaimed upon learning of my interest in evolution: 'But I thought Darwin had been disproved.' My mind started placing bets with itself: which particular second-hand, distorted half-truth has she misunderstood? I had just put my money on garbled Stephen Gould with a small side bet on (no need to garble) Fred Hoyle, when my companion revealed the winner as an older favourite: 'I heard that the missing link had now been shown to be a hoax.' Piltdown, by God, still raising his ugly cranium after all these years!

Incidents like this reveal the extreme flimsiness of the straws that will be clutched by those with a strong desire to believe something silly. There are between 3 and 30 million species alive today, and as many as a billion have probably existed since life began. Just one fossil of just one of those millions of species turns out to be a hoax. Yet of all the volumes and volumes of facts about evolution, the only thing that stuck in my companion's head was Piltdown. A parallel case is the extraordinarily popular aggrandizement of Eldredge and Gould's theory of 'punctuated equilibrium'. A minor dispute among experts (about whether evolution is smoothly continuous or interrupted by periods of stagnation when no evolutionary change occurs in a given lineage) has been blown up to give the impression that Darwinism's foundations are quivering. It is as if the discovery that the Earth is not a perfect sphere but an oblate spheroid cast sensational doubt on the whole Copernican world view, and reinstated flat-earthism. The anti-Darwinian sounding rhetoric of the punctuated equilibrists was a regrettable gift to creationists. Dr Gould regrets this as strongly as anyone, but I fear his protestations that his words have been misinterpreted will be to little avail.*

Whether Gould really has anything to answer for, he certainly has fought the good fight in the bizarre tragicomedy or tragifarce of modern American evolution politics. He travelled to Arkansas in 1981 to lend his formidable voice to the right side in the 'Scopes Trial II'. His obsession with history even took him on a visit to Dayton, Tennessee, scene of that previous Southern farce, and the subject of one of the most sympathetic and charming of the essays in the present book. His analysis of the appeal of creationism is wise and should be read by intolerant Darwin-freaks like me.

Gould's tolerance is his greatest virtue as a historian: that and

---

*'Since we proposed punctuated equilibria to explain trends, it is infuriating to be quoted again and again by creationists – whether by design or stupidity, I do not know – as admitting that the fossil record includes no transitional forms. Transitional groups are generally lacking at the species level but they are abundant between larger groups.' From the essay, 'Evolution as Fact and Theory', p. 260 of *Hen's Teeth and Horse's Toes*.

his warmth towards his subjects. His centennial tribute to Charles Darwin is offbeat in a characteristically delightful and affectionate way. Where others loftily pontificate, Gould goes down to earth and celebrates Darwin's last treatise, on worms. Darwin's worm book is not a 'harmless work of little importance by a great naturalist in his dotage'. It exemplifies his entire world view, based on the power of small causes, working together in large numbers and over long time spans, to wreak great changes:

> We who lack an appreciation of history and have so little feel for the aggregated importance of small but continuous change scarcely realize that the very ground is being swept from beneath our feet; it is alive and constantly churning ... Was Darwin really conscious of what he had done as he wrote his last professional lines, or did he proceed intuitively, as men of his genius sometimes do? Then I came to the last paragraph and I shook with the joy of insight. Clever old man; he knew full well. In his last words, he looked back to the beginning, compared those worms with his first corals and completed his life's work in both the large and the small ...

And the quotation of Darwin's last sentences follows.

Hen's Teeth and Horse's Toes is as enigmatic a title as *Pluto's Republic*, and it requires more explanation. If the present volume could be said to exercise a bee in Gould's bonnet, to distinguish it from its two predecessors, it is epitomized in the essay of the same name. I will explain the point rather fully, because it is one with which I strongly agree although I am supposed, apparently by Gould himself among others, to hold opposing views. I can sum the point up by giving a new twist to a phrase already twisted by Peter Medawar. If science is the art of the soluble, evolution is the art of the developable.

Development is change within an individual organism, from single cell to adult. Evolution is also change, but change of a type that requires subtler understanding. Each adult form in an evolutionary series will appear to 'change' into the next, but it is change only in the sense that each frame on a movie film 'changes' into the next. In reality, of course, each adult in the succession begins as a single cell and develops anew. Evolutionary change is change in genetically controlled processes of embryonic development, not literal change from adult form to adult form.

Gould fears that many evolutionists lose sight of development, and this leads them into error. There is firstly the error of genetic atomism, the fallacious belief in a one-to-one mapping between single genes and bits of body. Embryonic development doesn't work like that. The genome is not a 'blueprint'. Gould regards me as an arch genetic atomist, wrongly, as I have explained at length elsewhere.[117] It is one of those cases where you will misunderstand an author unless you interpret his words in the context of the position he was arguing against.

Consider the following, from Gould himself:

> Evolution is mosaic in character, proceeding at different rates in different structures. An animal's parts are largely dissociable, thus permitting historical change to proceed.

This appears to be rampant, and very un-Gouldian, atomism! Until you realize what Gould was arguing against: Cuvier's belief that evolution is impossible because change in any part is useless unless immediately accompanied by change in all other parts.* Similarly, the apparent genetic atomism that Gould criticizes in some other authors makes sense when you realize what those authors were arguing against: 'group selection' theories of evolution in which animals are supposed to act for the good of the species or some other large group. An atomistic interpretation of the role of genes in development is an error. An atomistic interpretation of the role of genetic differences in evolution is not an error, and is the basis of a telling argument against errors of the 'group selection' kind.

Atomism is just one of the errors that Gould sees as flowing from evolutionists' cavalier treatment of development. There are two others which are, on the face of it, opposite to each other: the error of assuming that evolution is too powerful, and the error of assuming that it is not powerful enough. The naive perfectionist believes that living material is infinitely ductile, ready to be shaped into whatever form natural selection dictates. This ignores

---

*A doctrine recently revived as 'irreducible complexity' under the mistaken impression that it is new.

the possibility that developmental processes are incapable of producing the desired form. The extreme 'gradualist' believes that all evolutionary changes are tiny, forgetting, according to Gould, that developmental processes can change in very large and complex ways, in single mutational steps. The general point, that we have to understand development before we can speculate constructively about evolution, is correct.

This must be what Medawar meant when he complained about 'the real weakness of modern evolutionary theory, namely its lack of a complete theory of variation, of the origin of candidature for evolution'. And this is why Gould is interested in hens' teeth and horses' toes. He makes the point that atavistic 'throwbacks', like hens with teeth and horses with three toes rather than one, are interesting because they tell us about the magnitude of evolutionary change that development allows. For the same reason he is interested in (and very interesting on) the development of zebras' stripes, and macromutations like insects with supernumerary thoraxes and wings.

I said that Gould and I were supposed to be professional adversaries and I would be disingenuous to pretend to like everything in this book. Why, for instance, does he find it necessary, after the phrase 'A strict Darwinian', to add '– I am not one –'? Of course Gould is a strict Darwinian, or if he isn't, nobody is; if you interpret 'strict' strictly enough, nobody is a strict anything. It is a pity, too, that Gould is still preaching against innocuous phrases like 'adultery in mountain bluebirds' and 'slavery in ants'. His rhetorical question about his own disapproval of such harmless anthropomorphisms, 'Is this not mere pedantic grousing', should be answered with a resounding 'Yes'. Gould himself made unselfconscious use of 'slavery of ants' in his own account of the phenomenon (*Ever Since Darwin*; presumably this was written in the days before some pompous comrade spotted the dangerous ideological implications of the phrase). Since our language grew in a human setting, if biologists tried to ban human imagery they would almost have to stop communicating. Gould is an expert communicator, and of course he in practice treats his own puritanical strictures with the contempt that he secretly knows they deserve. The very first essay of the present book tells us how two angler fish (*angler* fish?) are caught '*in flagrante delicto*' and discover 'for themselves what, according to Shakespeare, "every wise man's son doth know" – "journeys end in lovers meeting" '.

This is indeed a beautiful book, the pages glowing with a naturalist's love of life and a historian's respect and affection for his subjects, the vision extended and clarified by a geologist's familiarity with 'deep time'. To borrow a Medawarian phrase and like Peter Medawar himself, Stephen Gould is an aristocrat of learning. These are both extraordinarily gifted men, with some of the arrogance natural to aristocrats and those who have always been top of every class of which they have been members, but big enough to get away with it and generous enough to rise above arrogance too. Read their books if you are a scientist and, especially, read them if you are not.

5.3

Hallucigenia, *Wiwaxia* and Friends[118]

Review of *Wonderful Life* by S. J. Gould

Wonderful Life is a beautifully written and deeply muddled book. To make unputdownable an intricate, technical account of the anatomies of worms, and other inconspicuous denizens of a half-billion-year-old sea, is a literary *tour de force*. But the theory that Stephen Gould wrings out of his fossils is a sorry mess.

The Burgess Shale, a Canadian rock formation dating from the Cambrian, the earliest of the great fossil eras, is a zoological treasury. Freak conditions preserved whole animals, soft parts and all, in full 3-D. You can literally dissect your way through a 530-million-year-old animal. C. D. Walcott, the eminent palaeontologist who discovered the Burgess fossils in 1909, classified them according to the fashion of his time: he 'shoehorned' them all into modern groups. 'Shoehorn' is Gould's own excellent coining. It recalls to me my undergraduate impatience with a tutor who asked whether the vertebrates were descended from this invertebrate group or that. 'Can't you see,' I almost shouted, 'that our categories are all modern? Back in the Precambrian, we wouldn't have recognized those invertebrate groups anyway. You are asking a non-question.' My tutor agreed, and then went right on tracing modern animals back to other modern groups.

That was shoehorning, and that is what Walcott did to the Burgess animals. In the 1970s and 80s, a group of Cambridge palaeontologists returned to Walcott's museum specimens (with some newer collections from the Burgess site), dissected their 3-dimensional structure and overturned his classifications. These revisionists, principally Harry Whittington, Derek Briggs and Simon Conway Morris, are the heroes of Gould's tale. He milks every ounce of drama from their rebellion against the shoehorn, and at times he goes right over the top: 'I believe that

Whittington's reconstruction of *Opabinia* in 1975 will stand as one of the great documents in the history of human knowledge.'

Whittington and his colleagues realized that most of their specimens were far less like modern animals than Walcott had alleged. By the end of their epic series of monographs they thought nothing of coining a new phylum for a single specimen ('phylum' is the highest unit of zoological classification; even the vertebrates constitute only a sub-category of the Phylum Chordata). These brilliant revisions are almost certainly broadly correct, and they delight me beyond my undergraduate dreams. What is wrong is Gould's usage of them. He concludes that the Burgess fauna was demonstrably more diverse than that of the entire planet today, he alleges that his conclusion is deeply shocking to other evolutionists, and he thinks that he has upset our established view of history. He is unconvincing on the first count, clearly wrong on the second two.

In 1958 the palaeontologist James Brough published the following remarkable argument: evolution must have been qualitatively different in the earliest geological eras, because then new phyla were coming into existence; today only new species arise! The fallacy is glaring: every new phylum has to start as a new species. Brough was wielding the other end of Walcott's shoehorn, viewing ancient animals with the misplaced hindsight of a modern zoologist: animals that in truth were probably close cousins were dragooned into separate phyla because they shared key diagnostic features with their more divergent modern descendants. Gould too, even if he is not exactly reviving Brough's claim, is hoist with his own shoehorn.

How should Gould properly back up his claim that the Burgess fauna is super-diverse? He should – it would be the work of many years and might never be made convincing – take his ruler to the animals themselves, unprejudiced by modern preconceptions about 'fundamental body plans' and classification. The true index of how unalike two animals are is how unalike they actually are. Gould prefers to ask whether they are members of known phyla. But known phyla are modern constructions. Relative resemblance to modern animals is not a sensible way of judging how far Cambrian animals resemble one another.

The five-eyed, nozzle-toting *Opabinia* cannot be assimilated to any textbook phylum. But, since textbooks are written with modern animals in mind, this does not mean that *Opabinia* was, in fact, as different from its contemporaries as the status 'separate phylum' would suggest. Gould makes a token attempt to counter this criticism, but he is hamstrung by dyed-in-the-wool essentialism and Platonic ideal forms. He really seems unable to comprehend that animals are continuously variable functional machines. It is as though he sees the great phyla not diverging from early blood brothers but springing into existence fully differentiated.

Gould, then, singularly fails to establish his super-diversity thesis. Even if he were right, what would this tell us about 'the nature of history'? Since, for Gould, the Cambrian was peopled with a greater cast of phyla than now exist, we must be wonderfully lucky survivors. It could have been our ancestors who went extinct; instead it was Conway Morris's 'weird wonders', *Hallucigenia*, *Wiwaxia* and their friends. We came 'that close' to not being here.

Gould expects us to be surprised. Why? The view that he is attacking – that evolution marches inexorably towards a pinnacle such as man – has not been believed for years. But his quixotic strawmandering, his shameless windmill-tilting, seem almost designed to encourage misunderstanding (not for the first time: on a previous occasion he went so far as to write that the neo-Darwinian synthesis was 'effectively dead'). The following is typical of the publicity surrounding *Wonderful Life* (incidentally, I suspect that the lead sentence was added without the knowledge of the credited journalist): 'The human race did not result from the "survival of the fittest", according to the eminent American professor, Stephen Jay Gould. It was a happy accident that created Mankind.'[119] Such twaddle, of course, is nowhere to be found in Gould, but whether or not he seeks that kind of publicity, he all too frequently attracts it. Readers regularly gain the impression that he is saying something far more radical and surprising than he actually is.

Survival of the fittest means individual survival, not survival of major lineages. Any orthodox Darwinian would be entirely happy with major extinctions being largely a matter of luck. Admittedly there is a minority of evolutionists who think that Darwinian selection chooses between higher-level groupings. They are the only Darwinians likely to be disconcerted by Gould's 'contingent extinction'. And who is the most prominent advocate of higher-level selection today? You've guessed it. Hoist again!

Human Chauvinism and
Evolutionary Progress[120]

Review of *Full House* by S. J. Gould

This pleasantly written book has two related themes. The first is a statistical argument which Gould believes has great generality, uniting baseball, a moving personal response to the serious illness from which, thankfully, the author has now recovered, and his second theme: that of whether evolution is progressive. The argument about evolution and progress is interesting – though flawed as I shall show – and will occupy most of this review. The general statistical argument is correct and mildly interesting, but no more so than several other homilies of routine methodology about which one could sensibly get a bee in one's bonnet.

Gould's modest and uncontroversial statistical point is simply this. An apparent trend in some measurement may signify nothing more than a change in variance, often coupled with a ceiling or floor effect. Modern baseball players no longer hit a 0.400 (whatever that might be – evidently it is something pretty good). But this doesn't mean they are getting worse. Actually everything about the game is getting better and the variance is getting less. The extremes are being squeezed and 0.400 hitting, being an extreme, is a casualty. The apparent decrease in batting success is a statistical artefact, and similar artefacts dog generalizations in less frivolous fields.

That didn't take long to explain, but baseball occupies 55 jargon-ridden pages of this otherwise lucid book and I must enter a mild protest on behalf of those readers who live in that obscure and little known region called the rest of the world. I invite Americans to imagine that I spun out a whole chapter in the following vein:

> The home keeper was on a pair, vulnerable to anything from a yorker to a chinaman, when he fell to a googly given plenty of air. Silly mid on appealed for leg before, Dicky Bird's finger shot up and the tail collapsed. Not surprisingly, the skipper took the light. Next morning the night watchman, defiantly out of his popping crease, snicked a cover drive off a no ball straight through the gullies and on a fast outfield third man failed to stop the boundary … etc. etc.

Readers in England, the West Indies, Australia, New Zealand, India, Pakistan, Sri Lanka and anglophone Africa would understand every word, but Americans, after enduring a page or two, would rightly protest.

Gould's obsession with baseball is harmless and, in the small doses to which we have hitherto been accustomed, slightly endearing. But this hubristic presumption to sustain readers' attention through six chapters of solid baseball chatter amounts to American chauvinism (and I suspect American male chauvinism at that). It is the sort of self-indulgence from which an author should have been saved by editor and friends before publication – and for all I know they tried. Gould is normally so civilized in his cosmopolitan urbanity, so genial in wit, so deft in style. This book has a delightfully cultivated yet unpretentious 'Epilog on Human Culture' which I gratefully recommend to anyone, of any nation. He is so good at explaining science without jargon yet without talking down, so courteous in his judgement of when to spell out, when to flatter the reader by leaving just a little unsaid. Why does his gracious instinct desert him when baseball is in the air?

Another minor plaint from over the water, this time something which is surely not Dr Gould's fault: may I deplore the growing publishers' habit of gratuitously renaming books when they cross the Atlantic (both ways)? Two of my colleagues are at risk of having their (excellent, and already well-named) books retitled, respectively, 'The Pelican's Breast' and 'The Pony Fish's Glow' (now what, I wonder, can have inspired such flights of derivative imagination?) As one embattled author wrote to me, 'Changing the title is something big and important they can do to justify their salaries, and it does not require reading the book, so that's why they like it so much.' In the case of the book under review, if the author's own title, *Full House*, is good enough for the American market, why is the British edition masquerading under the alias of *Life's Grandeur*? Are we supposed to need protection from the argot of the card table?

At the best of times such title changes are confusing and mess up our literature citations. This particular change is doubly unfortunate because *Life's Grandeur* (the title, not the book) is tailor-made for confusion with *Wonderful Life*, and nothing about the difference between the titles conveys the difference between the contents. The two books are not Tweedledum and Tweedledee, and it is unfair on their author to label them as if they were. More generally, may I suggest that authors of the world unite and assert their right to name their own books.

Enough of carping. To evolution: is it progressive? Gould's definition of progress is a human-chauvinistic one which makes it all too easy to deny progress in evolution. I shall show that if we use a less anthropocentric, more biologically sensible, more 'adaptationist' definition, evolution turns out to be clearly and importantly progressive in the short to medium term. In another sense it is probably progressive in the long term too.

Gould's definition of progress, calculated to deliver a negative answer to the question whether evolution is progressive, is

> a tendency for life to increase in anatomical complexity, or neurological elaboration, or size and flexibility of behavioral repertoire, or any criterion obviously concocted (if we would only be honest and introspective enough about our motives) to place *Homo sapiens* atop a supposed heap.

My alternative, 'adaptationist' definition of progress is

> a tendency for lineages to improve cumulatively their adaptive fit to their particular way of life, by increasing the numbers of features which combine together in adaptive complexes.

I'll defend this definition and my consequent, limited, progressivist conclusion, later.

Gould is certainly right that human chauvinism, as an unspoken motif, runs through a great deal of evolutionary writing.

He'll find even better examples if he looks at the comparative psychology literature, which is awash with snobbish and downright silly phrases like 'subhuman primates', 'subprimate mammals' and 'submammalian vertebrates', implying an unquestioned ladder of life defined so as to perch us smugly on the top rung. Uncritical authors regularly move 'up' or 'down' the 'evolutionary scale' (bear in mind that they are in fact moving sideways among modern animals, contemporary twigs dotted all around the tree of life). Students of comparative mentality unabashedly and ludicrously ask, 'How far *down* the animal kingdom does learning extend?' Volume 1 of Hyman's celebrated treatise on the invertebrates is entitled 'Protozoa *through* Ctenophora' (my emphasis) – as if the phyla exist along an ordinal scale such that everybody knows which groups sit 'between' Protozoa and Ctenophora. Unfortunately, all zoology students do know – we've all been taught the same groundless myth.[121]

This is bad stuff, and Gould could afford to attack it even more severely than he attacks his normal targets. Whereas I would do so on logical grounds, Gould prefers an empirical assault. He looks at the actual course of evolution and argues that such apparent progress as can in general be detected is artefactual (like the baseball statistic). Cope's rule of increased body size, for example, follows from a simple 'drunkard's walk' model. The distribution of possible sizes is confined by a left wall, a minimal size. A random walk from a beginning near the left wall has nowhere to go but up the size distribution. The mean size has pretty well got to increase, and it doesn't imply a driven evolutionary trend towards larger size.

As Gould convincingly argues, the effect is compounded by a human tendency to give undue weight to new arrivals on the geological scene. Textbook biological histories emphasize a progression of grades of organization. As each new grade arrives, there is temptation to forget that the previous grades haven't gone away. Illustrators abet the fallacy when they draw, as representative of each era, only the newcomers. Before a certain date there were no eucaryotes. The arrival of eucaryotes looks more progressive than it really was because of the failure to depict the persisting hordes of procaryotes. The same false impression is conveyed with each new arrival on the stage: vertebrates, large-brained animals, and so on. An era may be described as the 'Age of Xs' – as though the denizens of the previous 'Age' had been replaced rather than merely supplemented.

Gould drives his point home with an admirable section on bacteria. For most of history, he reminds us, our ancestors have been bacteria. Most organisms still are bacteria, and a case can be made that most contemporary biomass is bacterial. We eucaryotes, we large animals, we brainy animals, are a recent wart on the face of a biosphere which is still fundamentally, and predominantly, procaryotic. To the extent that average size/complexity/cell number/brain size has increased since the 'age of bacteria', this could be simply because the wall of possibilities constrains the drunkard from moving in any other direction. John Maynard Smith recognized this possibility but doubted it when he considered the matter in 1970.[122]

> The obvious and uninteresting explanation of the evolution of increasing complexity is that the first organisms were necessarily simple ... And if the first organisms were simple, evolutionary change could only be in the direction of complexity.

Maynard Smith suspected that there was more to be said than this 'obvious and uninteresting explanation', but he didn't go into detail. Perhaps he was thinking of what he later came to term *The Major Transitions in Evolution*, or what I called 'The Evolution of Evolvability' (see below).

Gould's empirical treatment follows McShea[123], whose definition of complexity is reminiscent of J. W. S. Pringle's[124]; also of Julian Huxley's[125] definition of 'individuality' as 'heterogeneity of parts'. Pringle called complexity an epistemological concept, meaning a measure applied to our description of something rather than to that something itself. A crab is morphologically more complex than a millipede because, if you wrote a pair of books describing each animal down to the same level of detail, the crab book would have a higher word-count than the millipede book. The millipede book would describe a typical segment then simply add that, with listed exceptions, the other segments are the same. The crab book would require a separate chapter for each segment and would therefore have a higher information content.* McShea applied a similar notion to the vertebral column, expressing complexity in terms of heterogeneity among vertebrae.

With his measure of complexity in place, McShea sought statistical evidence for any general tendency for it to increase in fossil lineages. He made a distinction between passive trends (Gould's statistical artefacts) and driven trends (a true bias towards increased complexity, presumably driven by natural selection). By Gould's enthusiastic account, he concluded that there is no general evidence that a statistical majority of evolutionary lineages show driven trends in the direction of increased complexity. Gould goes further, pointing out that since so many species are parasites and parasite lineages commonly favour decreased complexity, there may even be a statistical trend in the opposite direction to the one hypothesized.

Gould is sailing dangerously close to the windmill-tilting that he has previously made his personal art form. Why should any thoughtful Darwinian have expected a majority of lineages to increase in anatomical complexity? Certainly it is not clear that anybody inspired by adaptationist philosophy would. Admittedly people inspired by human vanity might (and historically Gould is right that many have fallen for this vice). Our human line happens to have specialized in complexity, especially of the nervous system, so it is only human that we should define progress as an increase in complexity or in braininess. Other species will see it differently, as Julian Huxley[126] pointed out in a piece of verse entitled Progress:

The Crab to Cancer junior gave advice:
'Know what you want, my son, and then proceed
Directly sideways. God has thus decreed –
Progress is lateral; let that suffice'.

Darwinian Tapeworms on the other hand
Agree that Progress is a loss of brain,
And all that makes it hard for worms to attain
The true Nirvana – peptic, pure and grand.

*See also 'The "Information Challenge"' (pp. 118–19).

Man too enjoys to omphaloscopize.
Himself as Navel of the Universe ...

The poetry is not great (I couldn't bear to copy out the ending), and there is a confusion of timescales between the crab verse (behavioural time) and the tapeworm verse (evolutionary time), but an important point lurks here. Gould uses a human-chauvinistic definition of progress, measuring it in terms of complexity. This was why he was able to use parasites as ammunition against progress. Huxley's tapeworms, using a parasite-centred definition of progress, see the point with opposite sign. A statistically minded swift would search in vain for evidence that a majority of evolutionary lineages show trends towards improved flying performance. Learned elephants, to borrow a pleasantry from Steven Pinker[127], would ruefully fail to uphold the comforting notion that progress, defined as a driven elongation of the nose, is manifested by a statistical majority of animal lineages.

This may seem a facetious point but that is far from my intention. On the contrary, it goes to the heart of my adaptationist definition of progress. This, to repeat, takes progress to mean an increase, not in complexity, intelligence or some other anthropocentric value, but in the accumulating number of features contributing towards whatever adaptation the lineage in question exemplifies. By this definition, adaptive evolution is not just incidentally progressive, it is deeply, dyed-in-the-wool, indispensably progressive. It is fundamentally necessary that it should be progressive if Darwinian natural selection is to perform the explanatory role in our world view that we require of it, and that it alone can perform. Here's why.

Creationists love Sir Fred Hoyle's vivid metaphor for his own misunderstanding of natural selection. It is as if a hurricane, blowing through a junkyard, had the good fortune to assemble a Boeing 747. Hoyle's point is about statistical improbability. Our answer, yours and mine and Stephen Gould's, is that natural selection is cumulative. There is a ratchet, such that small gains are saved. The hurricane doesn't spontaneously assemble the airliner in one go. Small improvements are added bit by bit. To change the metaphor, however daunting the sheer cliffs that the adaptive mountain first presents, graded ramps can be found the other side and the peak eventually scaled.* Adaptive evolution must be gradual and cumulative, not because the evidence supports it (though it does) but because nothing except gradual accumulation could, in principle, do the job of solving the 747 riddle. Even divine creation wouldn't help. Quite the contrary, since any entity complicated and intelligent enough to perform the creative role would itself be the ultimate 747. And for exactly the same reason the evolution of complex, many-parted adaptations must be progressive. Later descendants will have accumulated a larger number of components towards the adaptive combination than earlier ancestors.

The evolution of the vertebrate eye must have been progressive. Ancient ancestors had a very simple eye, containing only a few features good for seeing. We don't need evidence for this (although it is nice that it is there). It has to be true because the alternative – an initially complex eye, well-endowed with features good for seeing – pitches us right back to Hoyle country and the sheer cliff of improbability. There must be a ramp of step-by-step progress towards the modern, multifeatured descendant of that optical prototype. Of course, in this case, modern analogues of every step up the ramp can be found, working serviceably in dozens of eyes dotted independently around the animal kingdom. But even without these examples, we could be confident that there must have been a gradual, progressive increase in the number of features which an engineer would recognize as contributing towards optical quality. Without stirring from our armchair, we can see that it must be so.

Darwin himself understood this kind of argument clearly, which is why he was such a staunch gradualist. Incidentally, it is also why Gould is unjust when he implies, not in this book but in many other places, that Darwin was against the spirit of punctuationism. The theory of punctuated equilibrium itself is gradualist (by Gad it had better be) in the sense in which Darwin was a gradualist – the sense in which all sane evolutionists must

*This rather coy allusion to *Climbing Mount Improbable* seemed appropriate because, as explained in the Preamble to this section, the Editor of *Evolution* had simultaneously commissioned a review of that book from Dr Gould.

be gradualists, at least where complex adaptations are concerned. It is just that, if punctuationism is right, the progressive, gradualistic steps are compressed into a timeframe which the fossil record does not resolve. Gould admits this when pressed, but he isn't pressed often enough.

Mark Ridley quotes Darwin on orchids, in a letter to Asa Gray: 'It is impossible to imagine so many co-adaptations being formed all by a chance blow.' As Ridley[128] goes on, 'The evolution of complex organs had to be gradual because all the correct changes would not occur in a single large mutation.' And gradual, in this context, needs to mean progressive in my 'adaptationist' sense. The evolution of anything as complex as an advanced orchid was progressive. So was the evolution of echolocation in bats and river dolphins – progressive over many, many steps. So was the evolution of electrolocation in fish, and of skull dislocation in snakes for swallowing large prey. So was the evolution of the complex of adaptations that equips cheetahs to kill, and the corresponding complex that equips gazelles to escape.

Indeed, as Darwin again realized, although he did not use the phrase, one of the main driving forces of progressive evolution is the coevolutionary arms race, such as that between predators and their prey. Adaptation to the weather, to the inanimate vicissitudes of ice ages and droughts, may well not be progressive: just an aimless tracking of unprogressively meandering climatic variables. But adaptation to the biotic environment is likely to be progressive because enemies, unlike the weather, themselves evolve. The resulting positive feedback loop is a good explanation for driven progressive evolution, and the drive may be sustained for many successive generations. The participants in the race do not necessarily survive more successfully as time goes by – their 'partners' in the coevolutionary spiral see to that (the familiar Red Queen Effect). But the equipment for survival, on both sides, is improving as judged by engineering criteria. In hard-fought examples we may notice a progressive shift in resources from other parts of the animal's economy to service the arms race.[129] And in any case the improvement in equipment will normally be progressive. Another kind of positive feedback in evolution, if R. A. Fisher and his followers are right, results from sexual selection. Once again, progressive evolution is the expected consequence.

Progressive increase in morphological complexity is to be expected only in taxa whose way of life benefits from morphological complexity.

Progressive increase in brain size is to be expected only in animals where braininess is an advantage. This may, for all I know, constitute a minority of lineages. But what I do insist is that in a majority of evolutionary lineages there will be progressive evolution towards something. It won't, however, be the same thing in different lineages (this was the point about swifts and elephants). And there is no general reason to expect a majority of lineages to progress in the directions pioneered by our human line.

But have I now defined progress so generally as to make it a blandly useless word? I don't think so. To say that the evolution of the vertebrate eye was progressive is to say something quite strong and quite important. If you could lay out all the inter-mediate ancestors in chronological order you'd find that, first, for a majority of dimensions of measurement, the changes would be transitive over the whole sequence. That is, if A is ancestral to B which is ancestral to C, the direction of change from A to B is likely to be the same as the direction of change from B to C. Second, the number of successive steps over which progress is seen is likely to be large: the transitive series extends beyond A, B and C, far down the alphabet. Third, an engineer would judge the performance to have improved over the sequence. Fourth, the number of separate features combining and conspiring to improve performance would increase. Finally, this kind of progress really matters because it is the key to answering the Hoyle challenge. There will be exceptional reversals, for instance in the evolution of blind cave fish, where eyes degenerate because they are not used and are costly to make. And there will doubtless be periods of stasis where there is no evolution at all, progressive or otherwise.

To conclude this point, Gould is wrong to say that the appearance of progress in evolution is a statistical illusion. It does not result just from a change in variance as a baseball-style artefact. To be sure, complexity, braininess and other particular qualities dear to the human ego should not necessarily be expected to increase progressively in a majority of lineages – though it would be interesting if they did: the investigations of McShea, Jerison[130] and others are not a waste of time. But if you define progress less chauvinistically – if you let the animals bring their own definition – you will find progress, in a genuinely interesting sense of the word, nearly everywhere.

Now it is important to stress that, on this adaptationist view (unlike the 'evolution of evolvability' view to be discussed shortly), progressive evolution is to be expected only on the short to medium term. Coevolutionary arms races may last for millions of years, but probably not hundreds of millions. Over the very long timescale, asteroids and other catastrophes bring evolution to a dead stop, major taxa and entire radiations go extinct. Ecological vacuums are created, to be filled by new adaptive radiations driven by new ranges of arms races. The several arms races between carnivorous dinosaurs and their prey were later mirrored by a succession of analogous arms races between carnivorous mammals and their prey. Each of these successive and separate arms races powered sequences of evolution which were progressive in my sense. But there was no global progress over the hundreds of millions of years, only a sawtooth succession of small progresses terminated by extinctions. Nonetheless, the ramp phase of each sawtooth was properly and significantly progressive.

Ironically for such an eloquent foe of progress, Gould flirts with the idea that evolution itself changes over the long haul, but he puts it in a topsy-turvy way which has undoubtedly been widely misleading. It is more fully expounded in *Wonderful Life* but reprised in the present book. For Gould, evolution in the Cambrian was a different kind of process from evolution today. The Cambrian was a period of evolutionary 'experiment', evolutionary 'trial and error', evolutionary 'false starts'. It was a period of 'explosive' invention, before evolution stabilized into the humdrum process we see today. It was the fertile time when all the great 'fundamental body plans' were invented. Nowadays, evolution just tinkers with old body plans. Back in the Cambrian, new phyla and new classes arose. Nowadays we only get new species!

This may be a slight caricature of Gould's own considered position, but there is no doubt that the many American non-specialists who unfortunately, as Maynard Smith[131] wickedly observes, get their evolutionary knowledge almost entirely from Gould, have been deeply misled. Admittedly, what follows is an extreme example, but Daniel Dennett has recounted a conversation with a philosopher colleague who read *Wonderful Life* as arguing that the Cambrian phyla did not have a common ancestor – that they had sprung up as independently initiated life forms! When Dennett assured him that this was not Gould's claim, his colleague's response was, 'Well then, what is all the fuss about?'

Even some professional evolutionists have been inspired by Gould's rhetoric into committing some pretty remarkable solecisms. Leakey and Lewin's *The Sixth Extinction*[132] is an excellent book except for its Chapter 3, 'The Mainspring of Evolution', which is avowedly heavily influenced by Gould. The following quotations from that chapter could hardly be more embarrassingly explicit:

Why haven't new animal body plans continued to crawl out of the evolutionary cauldron during the past hundreds of millions of years?

In early Cambrian times, innovations at the phylum level survived because they faced little competition.

Below the level of the family, the Cambrian explosion produced relatively few species, whereas in the post-Permian a tremendous species diversity burgeoned. Above family level however, the post-Permian radiation faltered, with few new classes and no new phyla being generated. Evidently, the mainspring of evolution operated in both periods, but it propelled greater extreme experimentation in the Cambrian than in the post-Permian, and greater variations on existing themes in the post-Permian.

Hence, evolution in Cambrian organisms could take bigger leaps, including phylum-level leaps, while later on it would be more constrained, making only modest jumps, up to the class level.

It is as though a gardener looked at an old oak tree and remarked, wonderingly: 'Isn't it strange that no major new boughs have appeared on this tree recently. These days, all the new growth appears to be at the twig level!'

As it happens, molecular clock evidence indicates that the 'Cambrian Explosion' may never have happened. Far from the major phyla diverging from a point at the beginning of the Cambrian, Wray, Levinton and Shapiro[133] present evidence that the common ancestors of the major phyla are staggered through hundreds of millions of years back in the Precambrian. But never mind that. That is not the point I want to make. Even if there really was a Cambrian explosion such that all the major phyla diverged during a ten million-year period, this is no reason to think that Cambrian evolution was a qualitatively special kind of super-jumpy process. *Baupläne* don't drop out of a clear Platonic sky, they evolve step by step from predecessors, and they do so (I bet, and so would Gould if explicitly challenged) under approximately the same Darwinian rules as we see today.

'Phylum-level leaps' and 'modest jumps, up to the class level' are the sheerest nonsense. Jumps above the species level don't happen, and nobody who thinks about it for two minutes claims that they do. Even the great phyla, when they originally bifurcated one from another, were just pairs of new species, members of the same genus. Classes are species that diverged a very long time ago, and phyla are species that diverged an even longer time ago. Indeed it is a moot – and rather empty – question precisely when in the course of the step-by-step, gradual mutual divergence of, say, mollusc ancestors and annelid ancestors after the time when they were congeneric species, we should wish to say that the divergence had reached 'Bauplan' status. A good case could be made that The Bauplan is a myth, probably as pernicious as any of the myths that Stephen Gould has so ably combatted, but this one, in its modern form, is largely perpetuated by him.

I return, finally, to the 'evolution of evolvability' and a very real sense in which evolution itself may evolve, progressively, over a longer timescale than the individual ramps of the arms race sawtooth. Notwithstanding Gould's just scepticism over the tendency to label each era by its newest arrivals, there really is a good possibility that major innovations in embryological technique open up new vistas of evolutionary possibility and that these constitute genuinely progressive improvements.* The origin of the chromosome, of the bounded cell, of organized meiosis, diploidy and sex, of the eucaryotic cell, of multicellularity, of gastrulation, of molluscan torsion, of segmentation – each of these may have constituted a watershed event in the history of life. Not just in the normal Darwinian sense of assisting individuals to survive and reproduce, but watershed in the sense of boosting evolution itself in ways that seem entitled to the label progressive. It may well be that after, say, the invention of multicellularity, or the invention of segmentation, evolution was never the same again. In this sense there may be a one-way ratchet of progressive innovation in evolution.

For this reason over the long term, and because of the cumulative character of coevolutionary arms races over the shorter term, Gould's attempt to reduce all progress to a trivial, baseball-style artefact constitutes a surprising impoverishment, an uncharacteristic slight, an unwonted demeaning of the richness of evolutionary processes.

---

*This is the idea that I dubbed 'The Evolution of Evolvability' (in C. Langton (ed.), *Artificial Life* (Santa Fe, Addison Wesley, 1982)) and Maynard Smith and Szathmáry wrote a book about (J. Maynard Smith and E. Szathmáry, *The Major Transitions in Evolution* (Oxford, W. H. Freeman/Spektrum, 1995)).

Unfinished Correspondence
with a Darwinian Heavyweight

The following email correspondence was never completed and now, sadly, it never can be.

9 December 2001

Stephen Jay Gould

Harvard

Dear Steve

Recently I received an email from Phillip Johnson, founder of the so-called 'Intelligent Design' school of creationists, crowing in triumph because one of his colleagues, Jonathan Wells, had been invited to take part in a debate at Harvard. He included the text of his email on his 'Wedge of Truth' web site, in which he announced the Wells debate under the headline 'Wells Hits a Home Run at Harvard'.

http://www.arn.org/docs/pjweekly/pj_weekly_011202.htm

The 'Home Run' turns out to be NOT a resounding success by Wells in convincing the audience, NOR any kind of besting of his opponent (Stephen Palumbi, who tells me he agreed to take part, with great reluctance, only because somebody at Harvard had ALREADY invited Wells and it was too late to do anything about that). There is no suggestion that Wells did well in the debate, nor even any obvious interest in whether he did. No, the 'Home Run' was simply and solely the matter of being invited by Harvard in the first place. These people have no hope of convincing reputable scientists by their ridiculous arguments. Instead, what they seek is the oxygen of respectability. We give them this oxygen by the mere act of ENGAGING with them at all. They don't mind being beaten in argument. What matters is that we give them recognition by bothering to argue with them in public.

You convinced me of this years ago when I phoned you up (you have probably forgotten this) to ask your advice when I was invited to debate Duane P Gish.

Ever since that phone call, I have repeatedly cited you and refused to debate these people, not because I am afraid of 'losing' the debate, but because, as you said, just to appear on a platform with them is to lend them the respectability they crave. Whatever might be the outcome of the debate, the mere fact that it is staged at all suggests to ignorant bystanders that there must be something worth debating, on something like equal terms.

First, I am interested to know whether you still hold to this view, as I do. Second, I am proposing that you might consider uniting with me (no need to involve others) in signing a short letter, say to the *New York Review of Books*, explaining publicly why we do not debate creationists (including the 'Intelligent Design' euphemism for creationists) and encouraging other evolutionary biologists to follow suit.

Such a letter would have great impact precisely because there have been widely publicised differences, and even animosities, between us (differences which creationists, with extreme intellectual dishonesty, have not hesitated to exploit). And I would not suggest writing a long disquisition on the technical differences which remain between us. That would only confuse the issue, make it harder to agree on a final draft, and lessen the impact. I wouldn't even mention our differences. I suggest a brief letter to the editor, explaining why we do not engage with 'intelligent design' or any other species of creationists, and offering our letter as a model for others to cite in refusing such invitations in the future. We both have better things to do with our time than give it over to such nonsense. Having just reached my sixtieth birthday (we are almost exactly the same age) I feel this keenly.

All best wishes

Richard

11 December 2001

Dear Richard,

Excellent idea — I'd be delighted to join you (and I agree it should just be you and me as signatories). Will you try a draft and send it to me?

I agree. Short and to the point. And NYRB could be best place.

I didn't realize we were so close in age (you look so young). Time does become ever more precious.

With my very best wishes.
Steve

14 December 2001

Dear Editor

Like any flourishing science, the study of evolution has its internal controversies, as we both know. But no qualified scientist doubts that evolution is a fact, in the ordinarily accepted sense in which it is a fact that the Earth orbits the Sun. It is a fact that human beings are cousins to monkeys, kangaroos, jellyfish and bacteria. No reputable biologist doubts this. Nor do reputable theologians, from the Pope on. Unfortunately, many lay Americans do, including some frighteningly influential, powerful and, above all, well-financed ones.

We are continually invited to engage in public debates against creationists, including latter-day creationists disguised under the euphemism 'Intelligent Design Theorists'. We always refuse, for one overriding reason. If we may be allowed to spell this reason out publicly, we hope our letter may be helpful to other evolutionary scientists plagued by similar invitations.

The question of who would 'win' such a debate is not at issue. Winning is not what these people realistically aspire to. The *coup* they seek is simply the recognition of being allowed to share a platform with a real scientist in the first place. This will suggest to innocent bystanders that there must be material here that is genuinely worth debating, on something like equal terms.

At the moment of writing, the leading 'Intelligent Design' website reports a debate at Harvard under the banner headline 'Wells Hits a Home Run at Harvard'.[134] Jonathan Wells is a creationist, incidentally a long-time devotee of the Unification Church (the 'Moonies').*

---

*'Darwinism: Why I went for a Second Ph.D.' is Jonathan Wells's own testimony on the turning point of his life: 'Father's words, my studies, and my prayers convinced me that I should devote my life to destroying Darwinism, just as many of my fellow Unificationists had already devoted their lives to destroying Marxism.

He had a debate last month against Stephen Palumbi, Professor of Biology at Harvard University. 'Home Run' might seem to suggest that Reverends (sic) Wells scored some kind of victory over Professor Palumbi. Or at least that he made powerful points and his speech was well received. No such claim is made. It doesn't even seem to be of interest.

The 'Home Run' turns out to be simply the public demonstration at Harvard that, in the words of the web site's author, Phillip Johnson, 'This is the sort of debate that is now occurring in universities.' There was a victory, but it occurred long before the debate itself. The creationist scored his home run at the moment the invitation from Harvard landed on his doormat. It came, by the way, not from any biological, or indeed scientific, department, but from the Institute of Politics.

Phillip Johnson himself, founding father of the 'Intelligent Design' movement (not a biologist, nor a scientist of any kind, but a lawyer who became a mid-life born-again Christian), wrote, in a letter of 6 April 2001, which he copied to one of us:

> It isn't worth my while to debate every ambitious Darwinist who wants to try his hand at ridiculing the opposition, so my general policy is that Darwinists have to put a significant figure at risk before I will agree to a debate. That means specifically Dawkins or Gould, or someone of like stature and public visibility.

Well, we can condescend too, and we have the advantage that evolutionary scientists don't need the publicity such debates can bring. In the unlikely event that a significant argument should ever emerge from the ranks of creationism/'intelligent design', we will be happy to debate it. Meanwhile, we shall cultivate our evolutionary gardens, occasionally engaging in the more exacting and worthwhile task of debating each other. What we shall not do is abet creationists in their disreputable quest for free publicity and unearned academic respectability.

When Father chose me (along with about a dozen other seminary graduates) to enter a Ph.D. program in 1978, I welcomed the opportunity to prepare myself for battle.' ('Father', of course, is the Moonies' name for Rev Moon himself). http://www.tparents.org/Library/Unification/Talks/Wells/DARWIN.htm.

In all humility, we offer these thoughts to our colleagues who receive similar invitations to debate.

Stephen Jay Gould, Harvard University

Richard Dawkins, Oxford University

Unfortunately, Steve never got around to revising the letter, which therefore lacks the stylish panache which his dexterous touch would have lent it. I received one further email, apologizing for the delay and hoping to deal with the matter soon. The subsequent silence, I now realize, coincided with his last illness. I therefore offer my draft, imperfect as it is, in the hope that it may go some way towards conveying the message which I originally learned from him many years ago. I have not removed his name from the draft, but it should be clearly understood that its faults are mine alone.

To close this section on a note of such concord may seem puzzling. Given that Steve was as much a neo-Darwinist as I am, what did we disagree about? The major disagreement emerges clearly out of his last big book, *The Structure of Evolutionary Theory*,[135] which I had no opportunity to see until after his death. It is appropriate, therefore, to spell out that issue here, and it also, as it happens, forms a natural bridge to the next essay. The question under dispute is this: what is the role of genes in evolution? Is it, to use Gould's phrase, 'book-keeping or causation'?

Gould saw natural selection as operating on many levels in the hierarchy of life. Indeed it may, after a fashion, but I believe that such selection can have evolutionary *consequences* only when the entities selected consist of 'replicators'. A replicator is a unit of coded information, of high fidelity but occasionally mutable, with some *causal* power over its own fate. Genes are such entities. So, in principle, are memes, but they are not under discussion here. Biological natural selection, at whatever level we may see it, results in evolutionary effects only insofar as it gives rise to changes in gene frequencies in gene pools. Gould, however, saw genes only as 'book-keepers', passively tracking the changes going on at other levels. In my view, whatever else genes are, they must be more than book-keepers, otherwise natural selection cannot work. If a genetic change has no causal influence on bodies, or at least on *something* that natural selection can 'see', natural selection cannot favour or disfavour it. No evolutionary change will result.

Gould and I would agree that genes can be seen as a book in which is written the evolutionary history of a species. In *Unweaving the Rainbow* I called it 'The Genetic Book of the Dead'. But the book is written via the natural selection of randomly varying genes, chosen by virtue of their causal influence on bodies. Book-keeping is precisely the wrong metaphor, because it reverses the causal arrow, almost in Lamarckian fashion, and makes the genes passive recorders. I dealt with this in 1982 (*The Extended Phenotype*) in my distinction between 'active replicators' and 'passive replicators'. The point is also explained in David Barash's superb review of Gould's book.[136]

Book-keeping is perversely – and characteristically – a valuable metaphor precisely because it is so diametrically back to front. Not for the first time, the characteristic vividness and clarity of a Gouldian metaphor helps us to see vividly and clearly what is wrong with the Gouldian message – and how it needs to be reversed in order to get at the truth.

I hope this brief note will not be seen as taking advantage to get the last word. *The Structure of Evolutionary Theory* is such a massively powerful last word, it will keep us all busy replying to it for years. What a brilliant way for a scholar to go. I shall miss him.

6

THERE IS ALL AFRICA
AND HER PRODIGIES IN US

I am one of those (it includes most people who have ever spent time south of the Sahara) who think of Africa as a place of enchantment. For me it stems from faint but haunting childhood memories, coupled with the mature understanding that Africa is our ancestral home. These themes recur throughout this section, and they introduce **Ecology of Genes** (6.1), my Foreword to Harvey Croze and John Reader's *Pyramids of Life*. This book uses Africa as an illuminating case study in the principles of ecology, and I used the opportunity of the Foreword to think about the relationship between ecology and natural selection. This could be seen as a continuation of my argument in the afterword to the previous section.

In this book, and elsewhere, I have been unkind to a view favoured by some social anthropologists, the 'cultural relativism' that acknowledges the equal status of many kinds of truth, scientific truth having no privileged rank among them. If ever I could be converted to some form of relativism, it might be after reading Elspeth Huxley's remarkable epic of Kenya, *Red Strangers*. **Out of the Soul of Africa** (6.2) is the Foreword to the new paperback edition of her novel. I wrote an article for the *Financial Times*, pointing out that *Red Strangers* had been out of print for years and challenging any publisher to do something about it. The admirable Penguin did, and they reprinted my article as the Foreword.

I am now waiting for a scholar of literature to explain to me why *Red Strangers* is not rated one of the great novels of the twentieth century, the equal of a John Steinbeck except that Elspeth Huxley's imagery is Kikuyu rather than American.

Run like the eland ... Run, warriors, with feet like arrows and the hearts of lions; the lives and wealth of your fathers are yours to save ... Their thighs were straight as saplings, their features sharp as axes, their skins lighter than honey. His limbs began to quiver like the wings of a sunbird when its beak sucks honey ...

It is a virtuoso feat of identification with another culture. Not only does she succeed in getting herself inside a Kikuyu skin, she achieves the same feat for the reader. And she makes you cry.

I am slightly ashamed to admit that another book that brings me close to tears – of joy this time – is a children's book. Or is it a very grown-up book which happens to be written by children? It is hard to decide, which is part of its charm and also probably why it has been unaccountably ignored by book reviews editors – they just didn't know which shelf to put it on. *The Lion Children* is about a family of children who are English, but whose home is a set of tents in Botswana, where they radio-track wild lions and are schooled entirely by their mother in the bush. They have written a book about their utterly extraordinary life. Never mind whether there is a conventionally labelled shelf to put it on, just read it. **I Speak of Africa and Golden Joys** (6.3), my Foreword, is reproduced here.

Last in this section is a travel piece, which again takes up the two themes of Africa as our ancestral home and Africa as my personal birthplace and weaves them together in an autobiographical story of travel and personal inspiration. The title was changed by the *Sunday Times* to 'All Our Yesterdays', but Macbeth's world-weariness is exactly opposite to the mood of my piece, so I am reverting to my original title, **Heroes and Ancestors** (6.4). Now that I think about it, Heroes and Ancestors would have made another fine title for this whole collection.

6.1
Ecology of Genes[137]
Foreword to *Pyramids of Life*
by Harvey Croze and John Reader

Africa was my personal cradle. But I left when I was seven, too young to appreciate – indeed the fact was not then known – that Africa is also humanity's cradle. The fossils of our species' formative years are all from Africa, and molecular evidence suggests that the ancestors of all today's peoples stayed there until as recently as the last hundred thousand years or so. We have Africa in our blood and Africa has our bones. We are all Africans.

This alone makes the African ecosystem an object of singular fascination. It is the community that shaped us, the commonwealth of animals and plants in which we served our ecological apprenticeship. But even if it were not our home continent Africa would captivate us, as perhaps the last great refugium of Pleistocene ecologies. If you want a late glimpse of the Garden of Eden, forget Tigris and Euphrates and the dawn of agriculture. Go instead to the Serengeti or the Kalahari. Forget the Arcadia of the Greeks and the dreamtime of the outback, they are so recent. Whatever may have come down the mountain at Olympus or Sinai, or even Ayers Rock, look instead to Kilimanjaro, or down the Rift Valley towards the High Veldt. There is where we were designed to flourish.

The 'design' of all living things and their organs is, of course, an illusion; an exceedingly powerful illusion, fabricated by a suitably powerful process, Darwinian natural selection. There is a second illusion of design in nature, less compelling but still appealing, and it is in danger of being mistaken for the first. This is the apparent design of ecosystems. Where bodies have parts that intricately harmonize and regulate to keep them alive, ecosystems have species that appear to do something similar at a higher level. There are the primary producers that convert raw solar energy into a form that others can use. There are the herbivores that consume them to use it, and then make a tithe of it available for carnivores and so on up the food chain – pyramid, rather, for the laws of thermodynamics rule that only a tenth of each level's energy shall make it to the level above. Finally, there are scavengers that recycle the waste products to make them available again, and in the process clean up the world and stop it becoming a tip. Everything fits with everything else like jigsaw pieces meshing in a huge multidimensional puzzle, and – as the cliché goes – we meddle with the parts at the risk of destroying a priceless whole.

The temptation is to think that this second illusion is crafted by the same kind of process as the first: by a version of Darwinian selection, but at a higher level. According to this erroneous view, the ecosystems that survive are the ones whose parts – species – harmonize, just as the organisms that survive in conventional Darwinism are the ones whose parts – organs and cells – work harmoniously for their survival. I believe that this theory is false. Ecosystems, like organisms, do indeed seem harmoniously designed; and the appearance of design is indeed an illusion. But there the resemblance ends. It is a different kind of illusion, brought about by a different process. The best ecologists, such as Croze and Reader, understand this.

Darwinism enters into the process, but it does not jump levels. Genes still survive, or fail to survive, within the gene pools of species, by virtue of their effects upon the survival and reproduction of the individual organisms that contain them. The illusion of harmony at a higher level is an indirect consequence of differential individual reproduction. Within any one species of animals or plants, the individuals that survive best are the ones that can exploit the other animals and plants, bacteria and fungi that are already flourishing in the environment. As Adam Smith understood long ago, an illusion of harmony and real efficiency will emerge in an economy dominated by self-interest at a lower level. A well balanced ecosystem is an economy, not an adaptation.

Plants flourish for their own good, not for the good of herbivores. But because plants flourish, a niche for herbivores opens up, and they fill it. Grasses are said to benefit from being grazed. The truth is more interesting. No individual plant benefits from being grazed per se. But a plant that suffers only slightly when it is grazed outcompetes a rival plant that suffers more. So successful grasses have benefited indirectly from the presence of grazers. And of course grazers benefit from the presence of grasses. Grasslands therefore build up as harmonious communities of relatively compatible grasses and grazers. They seem to cooperate. In a sense they do, but it is a modest sense that must be cautiously understood and judiciously understated. The same is true of the other African communities expounded by Croze and Reader.

I have said that the illusion of harmony at the ecosystem level is its own kind of illusion, different from, and emphatically not to be confused with, the Darwinian illusion that produces each efficiently working body. But a closer look reveals that there is a similarity after all, one that goes deeper than the – admittedly interesting and more commonly stated – observation that an animal can also be seen as a community of symbiotic bacteria. Mainstream Darwinian selection is the differential survival of genes within gene pools. Genes survive if they build bodies that flourish in their normal environment. But the normal environment of a gene importantly includes the other genes (strictly, their consequences) in the gene pool of the species. Natural selection therefore favours those genes that cooperate harmoniously in the joint enterprise of building bodies within the species. I have called the genes 'selfish cooperators'. There turns out to be, after all, an affinity between the harmony of a body and the harmony of an ecosystem. There is an ecology of genes.

Out of the Soul of Africa[138]

Foreword to *Red Strangers* by Elspeth Huxley

Elspeth Huxley died in 1997 at the age of 90. Best known for her vivid African memoirs, she was also a considerable novelist who, in *Red Strangers*, achieved a scale that could fairly be called epic. It is the saga of a Kikuyu family spanning four generations, beginning before the coming to Kenya of the British ('red' strangers because sunburned), and ending with the birth of a new baby girl, christened Aeroplane by her father ('His wife, he thought, would never be able to pronounce such a difficult word; but educated people would know, and understand'). Its 400 pages are gripping, moving, historically and anthropologically illuminating, humanistically mind-opening ... and, lamentably, out of print.*

I had an unrealized youthful ambition to write a science fiction novel. It would follow an expedition to, say, Mars, but seen through the eyes (or whatever passed for eyes) of the native inhabitants. I wanted to manoeuvre my readers into an acceptance of Martian ways so comprehensive that they would see the invading humans as strange and foreign aliens. It is Elspeth Huxley's extraordinary achievement in the first half of *Red Strangers* to immerse her readers so thoroughly in Kikuyu ways and thought that, when the British finally appear on the scene, everything about them seems to us alien, occasionally downright ridiculous, though usually to be viewed with indulgent tolerance. It is the same indulgent amusement, indeed, as I remember we bestowed upon Africans during my own colonial childhood.

Mrs Huxley, in effect, skilfully transforms her readers into Kikuyu, opening our eyes to see Europeans, and their customs, as we have never seen them before. We become used to an economy

---

*No longer!

pegged to the Goat Standard, so when coins (first rupees and then shillings) are introduced, we marvel at the absurdity of a currency that does not automatically accrue with each breeding season. We come to accept a world in which every event has a supernatural, magical interpretation, and feel personally swindled when the statement, 'The rupees that I pay you can later be changed into goats', turns out to be literally untrue. When Kichui (all white men are referred to by their Kikuyu nicknames) gives orders that his fields should be manured, we realize that he is mad. Why else would a man try to lay a curse upon his own cattle? 'Matu could not believe his ears. To bury the dung of a cow was to bring death upon it, just as death, or at any rate severe sickness, would come to a man whose excreta were covered with earth ... He refused emphatically to obey the order.' And, such is Elspeth Huxley's skill, even I, despising as I do the fashionable nostrums of 'cultural relativism', find myself endorsing Matu's sturdy good sense.

We are led to marvel at the absurdity of European justice, which seems to care *which* of two brothers committed a murder:

> ... what does it matter? Are not Muthengi and I brothers? Whichever it was that held the sword, our father Waseru and other members of our clan must still pay the blood-price.

Unaccountably, there is no blood price, and Matu, having cheerfully confessed to Muthengi's crime, goes to prison, where he leads 'a strange, comfortless life whose purpose he could not divine'. Eventually he is released. He has served his time but, since he didn't realize he was doing time, the event is of no significance. On returning to his own village, far from being disgraced, he has gained prestige from his sojourn with the mysterious strangers, who obviously regard him highly enough to invite him to live in their own place.

The novel takes us through episodes that we recognize as if from a great distance; through the First World War and the ravages of the subsequent Spanish flu, through smallpox epidemics and worldwide economic recession; and we never once are told in European terms that this is what is going on. We see all through Kikuyu eyes. The Germans are just another white tribe, and when the war ends we find ourselves wondering where are the plundered cattle that the victors ought to be driving home. What else, after all, is warfare *for?*

Ever since borrowing *Red Strangers* from the library, I have been on a ceaseless quest to acquire a copy of my own. It has been my routine first question on every visit to second-hand book fairs. Finally, I tracked down two old American copies simultaneously on the Internet. After so many years of restless searching, I could not resist buying both. So now, if any reputable publisher sincerely wants to look at *Red Strangers* with a view to bringing out a new edition,* I will gladly make available one of my hard-won copies. Nothing will part me from the other one.

*This was first published in the *Financial Times*. I am delighted to say that Penguin Books rose to this challenge and published the book, using my *Financial Times* article, here reproduced again, as the Foreword.

I Speak of Africa and Golden Joys[139]

Foreword to *The Lion Children*
by Angus, Maisie and Travers McNeice

This is an astonishing book, by an even more astonishing trio of children. It's hard to describe: you have to read it, and once you start reading you can't stop. Think of *Swallows and Amazons*, except that this story is true and it all happens far from the comfort of England. Think of *The Lion, the Witch and the Wardrobe*, except that the Lion Children need no magic wardrobe to pass through; no fake world of wonder. The real Africa, humanity's cradle, is more magical than anything C. S. Lewis could dream up. And, while they have no witch, these young authors do have a most remarkable mother. More of her in a moment.

Travers, Angus, Maisie and family have lived under canvas for almost as long as their little brother Oakley (think of *Just William*) can remember. All three of them have been driving Land Rovers ever since their feet could reach the pedals, and changing tyres (frequently) for as long as they've been strong enough to lift them.* They are self-sufficient and trustworthy far beyond their years, yet not in that disagreeable sense of being streetwise and fly. Field Marshall Montgomery once described Mao Tse Tung as the sort of man you could go into the jungle with. Well, I'm not sure I'd go with Mao Tse Tung into Hyde Park, but I would un-hesitatingly go into the jungle with Travers, Angus and Maisie, and no adult companions at all. No gun, just quick-witted young people with clear eyes, fast reflexes and most of a lifetime (albeit a short one) of African know-how. I don't know what to do if I meet an elephant. They do. I'm terrified of puff adders, mambas and scorpions. They take them in their stride. At the same time,

---

*Travers, Angus and Maisie were aged 16, 14 and 12 when they finished the book.

dependable and strong as they are, they still bubble with the innocence and charm of youth. This is still *Swallows and Amazons*, still an idyll, the sort of childhood that for most of us exists only in dreams and idealized misrememberings, 'the land of lost content'. Yet it is firmly in the real world. These innocents have seen favourite lions brutally killed, have rapped out reports of such tragedies in the dispassionate argot of the radio link, have assisted at the subsequent postmortems.

This accomplished book is entirely the work of its young authors, but it isn't hard to guess the source of their *ability* to do it – their imagination, their enterprise, their unorthodoxy, their adventurous spirit. My wife and I first met Kate Nicholls, their mother, in 1992 when she was living in the Cotswolds, pregnant with Oakley, commuting to study in Oxford libraries. A successful actress, she had become disillusioned with the stage and developed, in her late thirties, a passion (passion is the story of her life) for the science of evolution. Kate doesn't do anything by halves and, for her, an interest in evolution meant deep immersion in libraries, digging up the original research literature. With only minimal guidance from me in what became a series of informal tutorials, her reading transformed her into something of a scholarly authority on Darwinian theory. Her eventual decision to pull up roots and head for Botswana, where Darwinism can be daily witnessed in *practice*, seemed entirely in character: a natural, if unconventional, extension of the same scholarly quest. Her children, one can't help feeling, have a pretty fortunate inheritance, as well as an almost unique environment in which to realize it.

They also have to thank their mother for their education, and this is perhaps the most surprising aspect of their life. Quite soon after arriving in Botswana, Kate decided to teach them herself. A brave decision, I think I would have counselled against it. But I would have been wrong. Although all their schooling is done in camp, they keep proper terms, have challenging homework assignments and work towards internationally accredited exams. Kate gets good results by standard educational certifications, while at the same time tending, indeed enhancing, the natural sense of wonder that normal children too often lose during their teens. I don't think any reader of these pages could fail to judge her unorthodox School in the Bush a brilliant success. The proof lies in the book, for, to repeat, the children, and they alone, wrote it. All three authors show themselves to be excellent writers: sensitive, literate, articulate, intelligent and creative.

Kate's choice of Botswana rather than anywhere else in Africa was fortuitous. In the fullness of time it led to her meeting Pieter Kat. And of course the lions – wild lions, living and dying in the world for which the natural selection of their ancestors had prepared them. Pieter is the ideal stepfather for her children, and these young scientists have in turn become an indispensable part of the lion research and conservation project.

It wasn't till last year that my family and I finally visited the camp. The experience was unforgettable, and I can testify to the picture painted in *The Lion Children*. It really is just like that: more wonderful than mad, but a bit of both. My daughter Juliet went out ahead, part of a large invasion of young visitors who soon picked up the enthusiasm of the resident family. On Juliet's first full day in Africa, Travers took her out in a Land Rover, tracking radio-collared lions. When we received Juliet's letter home, brimming with excitement at such an initiation, I relayed the story to her grandmother, who interrupted me with panic in her voice: 'Plus, of course, at least two armed African rangers?' I had to confess that Travers really had been Juliet's only companion, that he had been driving the Land Rover all by himself, and that as far as I knew the camp boasted neither African rangers nor arms. I don't mind admitting that, though I concealed it from my mother, I was pretty anxious about the story myself. But that was before I had seen Travers in the bush. Or, indeed, Angus or Maisie.

We arrived a month after Juliet, and our fears were soon put to rest. I had been to Africa before, indeed was born there. But I have never felt so close to the wild. Or so close to lions or any large wild animals. And there was the marvellous camaraderie of life in camp; laughter and argument in the dining tent, everybody shouting at once. I think of sleeping and waking amid the sounds of the African night, the untiring 'Work harder' of the Cape Turtle Dove, the insolently robust barking of the baboons, the distant – and sometimes not so distant – roaring of the prides. I think of Juliet's sixteenth birthday party timed for the full moon: a surreal scene of candlelit table standing proud and alone on open ground, miles from camp and indeed from anywhere else; of the catch in the throat as we watched the huge moon rising exactly on cue, first reflected in the shallow Jackal Pan and later picking out the spectral shapes of marauding hyenas – which had us hastily bundling the sleeping Oakley into the safety of the Land Rover. I think of our last night and a dozen lions, gnawing and growling on a recently killed zebra only just outside the camp. The atavistic emotions that this primitive night scene aroused – for, whatever our upbringing, our genes are African – haunt me still.

But I can't begin to do justice to this world which has been the setting for such an extraordinary childhood. I was only there for a week, and I am no doubt jaded with maturity. Read the book and experience, through watchful young eyes, all Africa – and her prodigies.

6.4

Heroes and Ancestors[140]

Earliest memories can build a private Eden, a lost garden to which there is no return. The name Mbagathi conjured myths in my mind. Early in the war my father was called from the colonial service in Nyasaland (now Malawi) to join the army in Kenya. My mother disobeyed instructions to stay behind in Nyasaland and drove with him, along rutted dust roads and over unmarked and fortunately unpoliced borders, to Kenya, where I was later born and lived till I was two. My earliest memory is of the two white-washed thatched huts which my parents built for us in a garden, near the small Mbagathi river with its footbridge where I once fell into the water. I have always dreamed of returning to the site of this unwitting baptism, not because there was anything remarkable about the place, but because my memory is void before it.

That garden with the two whitewashed huts was my infant Eden and the Mbagathi my personal river. But on a larger timescale Africa is Eden to us all, the ancestral garden whose Darwinian memories have been carved into our DNA over millions of years until our recent worldwide 'Out of Africa' diaspora. It was at least partly the search for roots, our species' ancestors and my own childhood garden, that took me back to Kenya in December 1994.

My wife Lalla happened to sit next to Richard Leakey at a lunch to launch his *The Origin of Humankind* [141] and by the end of the meal he had invited her (and me) to spend Christmas with his family in Kenya. Could there be a better beginning to a search for roots than a visit to the Leakey family on their home ground? We accepted gratefully. On the way we spent a few days with an old colleague, the economic ecologist Dr Michael Norton-Griffiths, and his wife Annie, in their house at Langata, near Nairobi. This paradise of bougainvillea and lush green gardens was marred only by the evident necessity for the Kenya equivalent of the burglar alarm – the armed askari, hired to patrol the garden at night by every householder who can afford the luxury.

I didn't know where to start in quest of my lost Mbagathi. I knew only that it was somewhere near greater Nairobi. That the city had expanded since 1943 was only too obvious. For all I could tell, my childhood garden might languish under a car park or an international hotel. At a neighbour's carol-singing party I cultivated the greyest and most wrinkled guests, seeking an old brain in which the name of Mrs Walter, the philanthropic owner of our garden, or of Grazebrooks, her house, might have lodged. Though intrigued at my quest, none could help. Then I discovered that the stream below the Norton-Griffiths' garden was named the Mbagathi River. There was a steep red soil track down the hill and I made a ritual pilgrimage. At the foot of the hill, not 200 yards from where we were living, was a small footbridge and I stood and sentimentally watched the villagers returning home from work over the Mbagathi River.

I don't know, and probably never shall, if this was 'my' bridge, but it probably was my Jordan, for rivers outlive human works. I never discovered my garden and I doubt if it survives. Human memory is frail, our traditions as erratic as Chinese whispers and largely false, written records crumble and in any case writing is only millennia old. If we want to follow our roots back through the millions of years we need more persistent race memories. Two exist, fossils and DNA – hardware and software. The fact that our species now has a hard history is partly to the credit of one family, the Leakeys: the late Louis Leakey, his wife Mary, their son Richard and his wife Meave. It was to Richard and Meave's holiday house at Lamu that we were going for Christmas.

The engagingly filthy town of Lamu, one of the strongholds of Islam bordering the Indian Ocean, lies on a sandy island close to the mangrove fringes of the coast. The imposing waterfront recalls Evelyn Waugh's Matodi in the first chapter of *Black Mischief*. Open stone drains, grey with suds, line streets too narrow for wheeled traffic, and heavily laden donkeys purposefully trot their un-supervised errands across the town. Skeletal cats sleep in patches of sun. Black-veiled women like crows walk obsequiously past men seated on doorsteps, talking the heat and the flies away. Every four hours the muezzins (nowadays they are recorded on cassette tapes concealed in the minarets) caterwaul for custom. Nothing disturbs the Marabou storks at their one-legged vigil round the abattoir.

The Leakeys are white Kenyans, not English, and they built their house in the Swahili style (this is native Swahili country, unlike most of Kenya, where the Swahili language is an intro-duced lingua franca spread by the Arab slave trade). It is a large, white, thankfully cool cathedral of a house, with an arched veranda, tiles and rush matting on the floor, no glass in the windows, no hot water in the pipes and no need for either. The whole upstairs floor, reached by irregularly cut outside steps, is a single flat area furnished only with rush mats, cushions and mattresses, completely open to the warm night winds and the bats diving past Orion. Above this airy space, raised high on stilts, is the unique Swahili roof, thatched with reeds on a lofty superstructure of palm logs, intricately lashed together with thongs.

Richard Leakey is a robust hero of a man, who actually lives up to the cliché, 'a big man in every sense of the word'. Like other big men he is loved by many, feared by some, and not over-preoccupied with the judgements of any. He lost both legs in a near fatal air crash in 1993, at the end of his rampantly successful years crusading against poachers. As Director of the Kenya Wildlife Service he transformed the previously demoralized rangers into a crack fighting army with modern weapons to match those of the poachers and, more importantly, with an *esprit de corps* and a will to hit back at them. In 1989 he persuaded President Moi to light a bonfire of more than 2000 seized tusks, a uniquely Leakeyan masterstroke of public relations which did much to destroy the ivory trade and save the elephant. But jealousies were aroused by his international prestige which helped raise funds for his depart-ment, money which other officials coveted. Hardest to forgive, he conspicuously proved it possible to run a big department in Kenya efficiently and without corruption. Leakey had to go, and he did. Coincidentally, his plane had unexplained engine failure, and now he swings along on two artificial legs (with a spare pair specially made for swimming with flippers). He again races his sailing boat with his wife and daughters for crew, he lost no time in regaining his pilot's licence, and his spirit will not be crushed.

If Richard Leakey is a hero, he is matched in elephant lore by that legendary and redoubtable couple Iain and Oria Douglas-Hamilton. Iain and I had been students of the great naturalist, Niko Tinbergen, at Oxford, as had Mike Norton-Griffiths. It was a long time since we had met, and the Douglas-Hamiltons invited Lalla and me to Lake Naivasha for the final part of our holiday. Son of a dynasty of warlike Scottish lairds and more recently ace aviators, daughter of equally swashbuckling Italian-French adventurers in Africa, Iain and Oria met romantically, lived dangerously, raised their baby daughters to play fearlessly among wild elephants, fought the ivory trade with words and the poachers with guns.

Oria's parents, explorers and elephant hunters in the 1930s, built Sirocco, the 'pink palace', a stunning monument to art deco stylishness on the shores of Lake Naivasha, where they settled to farm 3000 acres. They are now buried side by side in the garden, near the avenue of cypresses that they planted to remind themselves of Naples, framing Longonot in place of Vesuvius. When they died the place fell into disrepair for ten years until a determined Oria, against all economic advice, returned. The farm now thrives again, though no longer 3000 acres, and Sirocco itself is restored, and is as it must have been. Iain flies his tiny plane home every weekend from Nairobi, where he runs his newly formed charity, 'Save the Elephants'. The family were all at Sirocco for Christmas and we were to join them for New Year.

Our arrival was unforgettable. Music thumped through the open doors (Vangelis's score for the film *1492* – I later chose it for *Desert Island Discs*). After a characteristic Italian and African lunch for 20 guests, we looked out over the terrace at the small paddock where, 25 years before, uninvited and unexpected, Iain had landed his plane to the terrified incredulity of Oria's parents and their guests at a similarly grand luncheon party. At dawn the morning after this sensational entrance into her life, Oria had without hesitation taken off with Iain for the shores of Lake Manyara, where the young man had begun his now famous study of wild elephants, and they have been together ever since. Their story is told in their two books, the Arcadian *Among the Elephants* and the more sombre *Battle for the Elephants*.[142]

On the veranda, staring towards Mount Longonot, is the skull of Boadicea, giant matriarch of Manyara, mother or grandmother of so many of Iain's elephants, victim of the poaching holocaust, her skull devotedly strapped into the back seat of Iain's plane and flown to its final rest overlooking a peaceful garden. There are no elephants in the Naivasha area, so we were spared the notorious Douglas-Hamilton treatment whereby guests are taken out and scared witless. The following passage, from the book *The Tree where Man was Born*,[143] by the American travel writer Peter Matthiessen, is entirely typical:

'I don't think she's going to charge us', Iain whispered. But the moment the herd was safely past, Ophelia swung up onto the bank, and she had dispensed with threat display. There were no flared ears, no blaring, only an oncoming cow elephant, trunk held high, less than twenty yards away.

As I started to run, I recall cursing myself for having been there in the first place; my one chance was that the elephant would seize my friend instead of me. In hopelessness, or perhaps some instinct not to turn my back on a charging animal, I faced around again almost before I had set out, and was rewarded with one of the great sights of a lifetime. Douglas-Hamilton, unwilling to drop his apparatus, and knowing that flight was useless anyway, and doubtless cross that Ophelia had failed to act as he predicted, was making a last stand. As the elephant loomed over us, filling the coarse heat of noon with her dusty bulk, he flared his arms and waved his glittering contraption in her face, at the same time bellowing, 'Bugger off!' Taken aback, the dazzled Ophelia flared her ears and blared, but she had sidestepped, losing the initiative, and now, thrown off course, she swung away toward the river, trumpeting angrily over her shoulder.

From high on the bank came a great peal of laughter from Oria. Iain and I trudged up to lunch; there was very damned little to say.

The only flaw in our Naivasha holiday was an ugly rumour that a leopard had been snared on a neighbouring farm and was painfully dragging the snare somewhere in the area. Grown quiet with anger, Iain took down his gun (for a wounded leopard can be dangerous), called for the best Masai tracker on the farm, and we set off in an ancient Land Rover.

The plan was to find the leopard by tracking and by questioning witnesses, lure it into a trap, nurse it back to health and release it again on the farm. Knowing no Swahili, I could gauge the progress of Iain's cross-examinations only by facial expressions, tones of voice and Iain's occasional summaries for my benefit. We eventually found a young man who had seen the leopard, though he denied it at first. Iain whispered to me that such initial denials – baffling to my naive straightforwardness – were ritual and normal. Eventually, without for a moment acknowledging that he had changed his story, the youth would lead us to the scene. Sure enough he did, and there the Masai tracker spotted leopard hairs and a possible spoor. He bounded, doubled up, through the papyrus reeds, followed by Iain and me. Just when I thought we were hopelessly lost, we re-emerged at our starting point. The trail had gone cold.

By similarly roundabout verbal skirmishings we tracked down a more recent witness who led us to another clearing in the papyrus, and Iain decided that here was the best site for a trap. He telephoned the Kenya Wildlife Service and they came, within the day, with a large iron cage filling the back of a Land Rover. Its door was designed to clang shut when the bait of meat was tugged. At dead of night we lurched and bumped through the papyrus and hippo dung, camouflaged the trap with foliage, laid a trail of raw meat to its entrance, baited it with half a sheep and went to bed.

The next day, Lalla and I were due to return to Nairobi and we left with the trap still baited, having attracted nothing more substantial than a marsh mongoose. Iain flew us in his little plane, hopping over steaming volcanic hills and down lake-filled valleys, over zebras and (almost) under giraffes, scattering the dust and the goats of the Masai villages, skirting the Ngong hills to Nairobi. At Wilson Airport, we chanced to run into Meave Leakey. She has now largely taken over the running of the fossil-hunting work from Richard, and she offered to introduce us to our ancestors in the vaults of the Kenya National Museum. This rare privilege was arranged for next day, the morning of our departure for London.

The great archaeologist Schliemann 'gazed upon the face of Agamemnon'. Well, good, the mask of a Bronze Age chieftain is a fine thing to behold. But as Meave Leakey's guest I have gazed upon the face of KNM-ER 1470 (*Homo habilis*), who lived and died 20,000 centuries before the Bronze Age began ...

Each fossil is accompanied by a meticulously accurate cast which you are allowed to hold and turn over as you look at the priceless original. The Leakeys told us that their team was opening up a new site at Lake Turkana, with fossils 4 million years old, older than any hominids so far discovered. In the week that I write this, Meave and her colleagues have published in *Nature* the first harvest of this ancient stratum: a newly discovered species, *Australopithecus anamensis*, represented by a lower jaw and various other fragments. The new finds suggest that our ancestors were already walking upright 4 million years ago, surprisingly (to some) close to our split from the lineage of chimpanzees.*

The leopard, Iain later told us, never came to the trap. He had feared that it would not, for the evidence of the second witness suggested that, fatally hobbled by the snare, it was already near death from starvation. For me, the most memorable part of that leopard-tracking day was my conversation with the two black rangers from the Kenya Wildlife Service who brought the trap. I was deeply impressed by the efficiency, humanity and dedication of these men. They were not allowed to let me photograph their operation, and they seemed a little reserved until I mentioned the name of Dr Leakey, their former leader, now in the political wilderness. Their eyes immediately lit up. 'Oh, you know Richard Leakey? What a wonderful man, a magnificent man!' I asked them how the Kenya Wildlife Service was faring nowadays. 'Oh

*Even older fossils have been discovered since this was first written.

well, we soldier on. We do our best. But it is not the same. What a magnificent man!'

We went to Africa to find the past. We found heroes and inspiration for the future, too.

7

A PRAYER FOR MY DAUGHTER

This last section, its title borrowed from W. B. Yeats, has a single item: an open letter to my daughter, written when she was ten. For most of her childhood, I unhappily saw her only for short periods at a time, and it was not easy to talk about the important things of life. I had always been scrupulously careful to avoid the smallest suggestion of infant indoctrination, which I think is ultimately responsible for much of the evil in the world. Others, less close to her, showed no such scruples, which upset me, as I very much wanted her, as I want all children, to make up her own mind freely when she became old enough to do so. I would encourage her to think, without telling her *what* to think. When she reached the age of ten, I thought about writing her a long letter. But to send it out of the blue seemed oddly formal and forbidding.

Then an opportunity fortuitously arose. My literary agent John Brockman, with his wife and partner Katinka Matson, conceived the idea of editing a book of essays as a rite-of-passage gift for their son Max. They invited clients and friends to contribute essays of advice or inspiration for a young person starting life. The invitation spurred me into writing, as an open letter, the advice to my daughter which I had previously been shy to give. The book itself, *How Things Are*, changed its mission halfway through its compilation. It remained dedicated to Max, but the subtitle became *A Science Tool-kit for the Mind* and later contributors were not asked to write specifically for a young person.

Eight years down the road, the legal onset of Juliet's adulthood happened to fall during the preparation of this collection, and the book is dedicated to her as an eighteenth birthday present, with a father's love.

7.1

Good and Bad Reasons
for Believing[144]

Dear Juliet

Now that you are ten, I want to write to you about something that is important to me. Have you ever wondered how we know the things that we know? How do we know, for instance, that the stars, which look like tiny pinpricks in the sky, are really huge balls of fire like the Sun and very far away? And how do we know that the Earth is a smaller ball whirling round one of those stars, the Sun?

The answer to these questions is 'evidence'. Sometimes evidence means actually seeing (or hearing, feeling, smelling ...) that something is true. Astronauts have travelled far enough from the Earth to see with their own eyes that it is round. Sometimes our eyes need help. The 'evening star' looks like a bright twinkle in the sky but with a telescope you can see that it is a beautiful ball – the planet we call Venus. Something that you learn by direct seeing (or hearing or feeling ...) is called an observation.

Often evidence isn't just observation on its own, but observation always lies at the back of it. If there's been a murder, often nobody (except the murderer and the dead person!) actually observed it. But detectives can gather together lots of other observations which may all point towards a particular suspect. If a person's fingerprints match those found on a dagger, this is evidence that he touched it. It doesn't prove that he did the murder, but it can help when it's joined up with lots of other evidence. Sometimes a detective can think about a whole lot of observations and suddenly realize that they all fall into place and make sense if so-and-so did the murder.

Scientists – the specialists in discovering what is true about the world and the universe – often work like detectives. They make a guess (called a hypothesis) about what might be true. They then say to themselves: *if* that were really true, we ought to see so-and-so. This is called a prediction. For example, if the world is really round, we can predict that a traveller, going on and on in the same direction, should eventually find himself back where he started. When a doctor says that you have measles he doesn't take one look at you and *see* measles. His first look gives him a *hypothesis* that you *may* have measles. Then he says to himself: if she really has measles, I ought to see ... Then he runs through his list of predictions and tests them with his eyes (have you got spots?), his hands (is your forehead hot?), and his ears (does your chest wheeze in a measly way?). Only then does he make his decision and say, 'I diagnose that the child has measles.' Sometimes doctors need to do other tests like blood tests or X-rays, which help their eyes, hands and ears to make observations.

The way scientists use evidence to learn about the world is much cleverer and more complicated than I can say in a short letter. But now I want to move on from evidence, which is a good reason for believing something, and warn you against three bad reasons for believing anything. They are called 'tradition', 'authority' and 'revelation'.

First, tradition. A few months ago, I went on television to have a discussion with about 50 children. These children were invited because they'd been brought up in lots of different religions. Some had been brought up as Christians, others as Jews, Muslims, Hindus or Sikhs. The man with the microphone went from child to child, asking them what they believed. What they said shows up exactly what I mean by 'tradition'. Their beliefs turned out to have no connection with evidence. They just trotted out the beliefs of their parents and grandparents, which, in turn, were not based upon evidence either. They said things like, 'We Hindus believe so and so.' 'We Muslims believe such and such.' 'We Christians believe something else.'

Of course, since they all believed different things, they couldn't all be right. The man with the microphone seemed to think this quite proper, and he didn't even try to get them to argue out their differences with each other. But that isn't the point I want to make. I simply want to ask where their beliefs came from. They came from tradition. Tradition means beliefs handed down from grandparent to parent to child, and so on. Or from books handed down through the centuries. Traditional beliefs often start from almost nothing; perhaps somebody just makes them up originally, like the stories about Thor and Zeus. But after they've been handed down over some centuries, the mere fact that they are so old makes them seem special. People believe things simply because people have believed the same thing over centuries. That's tradition.

The trouble with tradition is that, no matter how long ago a story was made up, it is still exactly as true or untrue as the original story was. If you make up a story that isn't true, handing it down over any number of centuries doesn't make it any truer!

Most people in England have been baptized into the Church of England, but this is only one of many branches of the Christian religion. There are other branches such as the Russian Orthodox, the Roman Catholic and the Methodist churches. They all believe different things. The Jewish religion and the Muslim religion are a bit more different still; and there are different kinds of Jews and of Muslims. People who believe even slightly different things from each other often go to war over their disagreements. So you might think that they must have some pretty good reasons – evidence – for believing what they believe. But actually their different beliefs are entirely due to different traditions.

Let's talk about one particular tradition. Roman Catholics believe that Mary, the mother of Jesus, was so special that she didn't die but was lifted bodily into Heaven. Other Christian traditions disagree, saying that Mary did die like anybody else. These other religions don't talk about her much and, unlike Roman Catholics, they don't call her the 'Queen of Heaven'. The tradition that Mary's body was lifted into Heaven is not a very old one. The Bible says nothing about how or when she died; in fact the poor woman is scarcely mentioned in the Bible at all. The belief that her body was lifted into Heaven wasn't invented until about six centuries after Jesus's time. At first it was just made up, in the same way as any story like *Snow White* was made up. But, over the centuries, it grew into a tradition and people started to take it seriously simply *because* the story had been handed down over so many generations. The older the tradition became, the more people took it seriously. It finally was written down as an official Roman Catholic belief only very recently, in 1950. But the story was no more true in 1950 than it was when it was first invented 600 years after Mary's death.

I'll come back to tradition at the end of my letter, and look at it in another way. But first I must deal with the two other bad reasons for believing in anything: authority and revelation.

Authority, as a reason for believing something, means believing it because you are told to believe it by somebody important. In the Roman Catholic Church, the Pope is the most important person, and people believe he must be right just because he is the Pope. In one branch of the Muslim religion, the important people are old men with beards called Ayatollahs. Lots of young Muslims are prepared to commit murder, purely because the Ayatollahs in a faraway country tell them to.*

When I say that it was only in 1950 that Roman Catholics were finally told that they had to believe that Mary's body shot off to Heaven, what I mean is that in 1950 the Pope told people that they had to believe it. That was it. The Pope said it was true, so it had to be true! Now, probably some of the things that Pope said in his life were true and some were not true. There is no good reason why, just because he was the Pope, you should believe everything he said, any more than you believe everything that lots of other people say. The present Pope has ordered his followers not to limit the number of babies they have. If people follow his authority as slavishly as he would wish, the results could be terrible famines, diseases and wars, caused by overcrowding.

Of course, even in science, sometimes we haven't seen the evidence ourselves and we have to take somebody else's word for it. I haven't, with my own eyes, seen the evidence that light travels at a speed of 186,000 miles per second. Instead, I believe books that tell me the speed of light. This looks like 'authority'. But actually it is much better than authority because the people who wrote the books have seen the evidence and anyone is free to look carefully at the evidence whenever they want. That is very

*The fatwah against Salman Rushdie was prominently in the news at the time.

comforting. But not even the priests claim that there is any evidence for their story about Mary's body zooming off to Heaven.

The third kind of bad reason for believing anything is called 'revelation'. If you had asked the Pope in 1950 how he knew that Mary's body disappeared into Heaven, he would probably have said that it had been 'revealed' to him. He shut himself in his room and prayed for guidance. He thought and thought, all by himself, and he became more and more sure inside himself. When religious people just have a feeling inside themselves that something must be true, even though there is no evidence that it is true, they call their feeling 'revelation'. It isn't only popes who claim to have revelations. Lots of religious people do. It is one of their main reasons for believing the things that they do believe. But is it a good reason?

Suppose I told you that your dog was dead. You'd be very upset, and you'd probably say, 'Are you sure? How do you know? How did it happen?' Now suppose I answered: 'I don't actually know that Pepe is dead. I have no evidence. I just have this funny feeling deep inside me that he is dead.' You'd be pretty cross with me for scaring you, because you'd know that an inside 'feeling' on its own is not a good reason for believing that a whippet is dead. You need evidence. We all have inside feelings from time to time, and sometimes they turn out to be right and sometimes they don't. Anyway, different people have opposite feelings, so how are we to decide whose feeling is right? The only way to be sure that a dog is dead is to see him dead, or hear that his heart has stopped; or be told by somebody who has seen or heard some real evidence that he is dead.

People sometimes say that you must believe in feelings deep inside, otherwise you'd never be confident of things like 'My wife loves me'. But this is a bad argument. There can be plenty of evidence that somebody loves you. All through the day when you are with somebody who loves you, you see and hear lots of little titbits of evidence, and they all add up. It isn't a purely inside feeling, like the feeling that priests call revelation. There are outside things to back up the inside feeling: looks in the eye, tender notes in the voice, little favours and kindnesses; this is all real evidence.

Sometimes people have a strong inside feeling that somebody loves them when it is not based upon any evidence, and then they are likely to be completely wrong. There are people with a strong inside feeling that a famous film star loves them, when really the film star hasn't even met them. People like that are ill in their minds. Inside feelings must be backed up by evidence, otherwise you just can't trust them.

Inside feelings are valuable in science too, but only for giving you ideas that you later test by looking for evidence. A scientist can have a 'hunch' about an idea that just 'feels' right. In itself, this is not a good reason for believing something. But it can be a good reason for spending some time doing a particular experiment, or looking in a particular way for evidence. Scientists use inside feelings all the time to get ideas. But they are not worth anything until they are supported by evidence.

I promised that I'd come back to tradition, and look at it in another way. I want to try to explain why tradition is so important to us. All animals are built (by the process called evolution) to survive in the normal place in which their kind live. Lions are built to be good at surviving on the plains of Africa. Crayfish are built to be good at surviving in fresh water, while lobsters are built to be good at surviving in the salt sea. People are animals too, and we are built to be good at surviving in a world full of ... other people. Most of us don't hunt for our own food like lions or lobsters, we buy it from other people who have bought it from yet other people. We 'swim' through a 'sea of people'. Just as a fish needs gills to survive in water, people need brains that make them able to deal with other people. Just as the sea is full of salt water, the sea of people is full of difficult things to learn. Like language.

You speak English but your friend Ann-Kathrin speaks German. You each speak the language that fits you to 'swim about' in your own separate 'people sea'. Language is passed down by tradition. There is no other way. In England, Pepe is a dog. In Germany he is *ein Hund*. Neither of these words is more correct, or more true than the other. Both are simply handed down. In order to be good at 'swimming about in their people sea', children have to learn the language of their own country, and lots of other things about their own people; and this means that they have to absorb, like blotting paper, an enormous amount of traditional information. (Remember that traditional information just means things that are handed down from grandparents to parents to children.) The child's brain has to be a sucker for traditional information. And the child can't be expected to sort out good and useful traditional information, like the words of a language, from bad or silly traditional information, like believing in witches and devils and ever-living virgins.

It's a pity, but it can't help being the case, that because children have to be suckers for traditional information, they are likely to believe anything the grown-ups tell them, whether true or false, right or wrong. Lots of what the grown-ups tell them is true and based on evidence, or at least sensible. But if some of it is false, silly or even wicked, there is nothing to stop the children believing that too. Now, when the children grow up, what do they do? Well, of course, they tell it to the next generation of children. So, once something gets itself strongly believed – even if it is completely untrue and there never was any reason to believe it in the first place – it can go on forever.

Could this be what has happened with religions? Belief that there is a god or gods, belief in Heaven, belief that Mary never died, belief that Jesus never had a human father, belief that prayers are answered, belief that wine turns into blood – not one of these beliefs is backed up by any good evidence. Yet millions of people believe them. Perhaps this is because they were told to believe them when they were young enough to believe anything.

Millions of other people believe quite different things, because they were told different things when they were children. Muslim children are told different things from Christian children, and both grow up utterly convinced that they are right and the others are wrong. Even within Christians, Roman Catholics believe different things from Church of England people or Episcopalians, Shakers or Quakers, Mormons or Holy Rollers, and all are utterly convinced that they are right and the others are wrong. They believe different things for exactly the same kind of reason as you speak English and Ann-Kathrin speaks German. Both languages are, in their own country, the right language to speak. But it can't be true that different religions are right in their own countries, because different religions claim that opposite things are true. Mary can't be alive in the Catholic Republic but dead in Protestant Northern Ireland.

What can we do about all this? It is not easy for you to do anything, because you are only ten. But you could try this. Next time somebody tells you something that sounds important, think to yourself: 'Is this the kind of thing that people probably know because of evidence? Or is it the kind of thing that people only believe because of tradition, authority or revelation?' And, next time somebody tells you that something is true, why not say to them: 'What kind of evidence is there for that?' And if they can't give you a good answer, I hope you'll think very carefully before you believe a word they say.

Your loving
Daddy

ENDNOTES

1. http://www.e-fabre.net/virtual library/more hunting wasp/chap04.htm
2. G. C. Williams, *Plan & Purpose in Nature* (New York, Basic Books, 1996), p. 157
3. http://www.apologeticspress.org/bibbul/2001/bb-01-75.htm
4. *Anticipations of the reaction of mechanical and scientific progress upon human life and thought* (London, Chapman and Hall, 1902)
5. J. Huxley, *Essays of a Biologist* (London, Chatto & Windus, 1926)
6. http://aleph0.clarku.edu/huxley/CE9/E-E.html
7. R. Dawkins, *The Selfish Gene* (Oxford, Oxford University Press, 1976; 2nd edn 1989). R. Dawkins, *The Blind Watchmaker* (London, Longman, 1986; London, Penguin, 2000)
8. Huxley (1926), *ibid.*
9. J. Huxley, *Essays of a Humanist* (London, Penguin, 1966)
10. Theodosius Dobzhansky, 'Changing Man', *Science*, 155 (27 January 1967), 409
11. First published as 'Hall of Mirrors' in *Forbes ASAP*, 2 October 2000
12. Published in the UK as *Intellectual Impostures* (London, Profile Books, 1998). My review of this book is reprinted on page 55 as 'Postmodernism Disrobed'
13. P. Gross and N. Levitt, *Higher Superstition* (Baltimore, The Johns Hopkins University Press, 1994)
14. D. Patai and N. Koertge, *Professing Feminism: Cautionary Tales from the Strange World of Women's Studies* (New York, Basic Books, 1994)
15. R. Dawkins, *River Out of Eden* (New York, Basic Books, 1995)
16. This interpretation of illusions is the one offered by our greatest living authority on them, Richard Gregory, *Eye and Brain*, 5th edn (Oxford, Oxford University Press, 1998)
17. L. Wolpert, *The Unnatural Nature of Science* (London, Faber & Faber, 1993)
18. From P. Cavalieri and P. Singer (eds.), *The Great Ape Project* (London, Fourth Estate, 1993)
19. R. Dawkins, *Unweaving the Rainbow* (London, Allen Lane/Penguin Press, 1998)

20 First published in *The Observer*, 16 November 1997

21 First published in the *Sunday Telegraph*, 18 October 1998

22 Review of Alan Sokal and Jean Bricmont, *Intellectual Impostures* (London, Profile Books, 1998); published in the US as *Fashionable Nonsense* (New York, Picador USA, 1998). *Nature*, 394 (9 July 1998), 141–3

23 P. B. Medawar, *Pluto's Republic* (Oxford, Oxford University Press, 1982)

24 Originally published in *The Guardian*, 6 July 2002

25 H. G. Wells, *The Story of a Great Schoolmaster: being a plain account of the life and ideas of Sanderson of Oundle* (London, Chatto & Windus, 1924)

26 *Sanderson of Oundle* (London, Chatto & Windus, 1926)

27 Originally published as the Foreword to the Student Edition of *The Descent of Man* (London, Gibson Square Books, 2002)

28 'Letter to Wallace, 26 February 1867' in Francis Darwin (ed.), *Life and Letters of Charles Darwin*, vol. 3 (London, John Murray, 1888), p. 95

29 H. Cronin, *The Ant and the Peacock* (Cambridge, Cambridge University Press, 1991)

30 W. D. Hamilton, *Narrow Roads of Gene Land*, vol. 2 (Oxford, Oxford University Press, 2001)

31 A. Zahavi and A. Zahavi, *The Handicap Principle: a missing piece of Darwin's puzzle* (Oxford, Oxford University Press, 1997)

32 R. A. Fisher, *The Genetical Theory of Natural Selection* (Oxford, Clarendon Press, 1930)

33 My own attempt at explaining it constitutes Chapter 8 of *The Blind Watchmaker*. For an authoritative modern survey of sexual selection, see M. Andersson, *Sexual Selection* (Princeton, Princeton University Press, 1994)

34 W. G. Eberhard, *Sexual Selection and Animal Genitalia* (Cambridge, Mass., Harvard University Press, 1988)

35 D. Dennett, *Darwin's Dangerous Idea* (New York, Simon & Schuster, 1995)

36 M. Ghiselin, *The Triumph of the Darwinian Method* (Berkeley, University of California Press, 1969)

37 R. Dawkins, 'Higher and Lower Animals: a Diatribe' in E. Fox-Keller and E. Lloyd (eds.), *Keywords in evolutionary biology* (Cambridge, Mass., Harvard University Press, 1992)

38 Charles Darwin, *The Descent of Man*, chapter XX of 1st edn, chapter XIX of 2nd edn

39 http://members.shaw.ca/mcfetridge/darwin.html

40 http://www.workersliberty.org/wlmags/wl61/dawkins.htm

41 Fisher (1930), *ibid.*

42 Letter dated 'Tuesday, February, 1866'. Published in James Marchant, *Alfred Russel Wallace: Letters and Reminiscences*, vol. 1 (London, Cassell,

1916). Reproduced by courtesy of the British Library, thanks to Dr Jeremy John

43 Fisher (1930), *ibid.*

44 W. D. Hamilton, 'Extraordinary Sex Ratios' (1966). Reprinted in his *Narrow Roads of Gene Land*, vol. 1 (Oxford, W. H. Freeman, 1996)

45 E. L. Charnov, *The Theory of Sex Allocation* (Princeton, Princeton University Press, 1982)

46 A. W. F. Edwards, 'Natural Selection and the Sex Ratio: Fisher's Sources', *American Naturalist*, 151 (1998), 564–9

47 R. L. Trivers, 'Parental investment and sexual selection' in B. Campbell (ed.), *Sexual Selection and the Descent of Man* (Chicago, Aldine, 1972), pp. 136–79

48 R. Leakey, *The Origin of Humankind* (London, Weidenfeld & Nicolson, 1994)

49 S. Pinker, *The Language Instinct* (London, Penguin, 1994)

50 S. J. Gould, *Ontogeny and Phylogeny* (Cambridge, Mass., Harvard University Press, 1977)

51 J. Diamond, *The Rise and Fall of the Third Chimpanzee* (London, Radius, 1991)

52 D. Morris, *Dogs: The ultimate dictionary of over 1000 dog breeds* (London, Ebury Press, 2001)

53 C. Vilà, J. E. Maldonado and R. K. Wayne, 'Phylogenetic Relationships, Evolution, and Genetic Diversity of the Domestic Dog', *Journal of Heredity*, 90 (1999), 71–7

54 G. Miller, *The Mating Mind* (London, Heinemann, 2000)

55 From M. H. Robinson and L. Tiger (eds.), *Man and Beast Revisited* (Washington, Smithsonian Institution Press, 1991)

56 R. Dawkins, 'Universal Darwinism' in D. S. Bendall (ed.), *Evolution from Molecules to Men* (Cambridge, Cambridge University Press, 1983), pp. 403–25. R. Dawkins, *The Blind Watchmaker* (New York, W. W. Norton, 1986), Chapter 11

57 C. Singer, *A Short History of Biology* (Oxford, Clarendon Press, 1931)

58 W. Bateson, quoted in E. Mayr, *The Growth of Biological Thought: Diversity, Evolution, and Inheritance* (Cambridge, Mass., Harvard University Press, 1982)

59 G. C. Williams, *Adaptation and Natural Selection* (Princeton, Princeton University Press, 1966)

60 R. A. Fisher, *The Genetical Theory of Natural Selection* (Oxford, Clarendon Press, 1930)

61 Dawkins, *The Blind Watchmaker*, p. 31

62 Peter Atkins, *The Second Law* (New York, Scientific American Books, 1984), and *Galileo's Finger* (Oxford, Oxford University Press, 2003) are characteristically lucid

63 R. Dawkins, *Climbing Mount Improbable* (London, Penguin, 1996), chapter 3

64 E. Mayr, *The Growth of Biological Thought: Diversity, Evolution, and Inheritance* (Cambridge, Mass., Harvard University Press, 1982)

65 F. H. C. Crick, *Life itself* (London, Macdonald, 1982)

66 R. Dawkins, *The Extended Phenotype* (San Francisco, W. H. Freeman, 1982/ Oxford, Oxford University Press, 1999), pp. 174–6. See also Endnote 36 and Dawkins, *The Blind Watchmaker*, chapter 11

67 Originally published in *the Skeptic*, 18, No. 4, December 1998 (Sydney, Australia)

68 Originally published in the *Daily Telegraph*, 17 July 1993, under the title 'Don't panic; take comfort, it's not all in the genes'

69 D. H. Hamer *et al.*, 'A linkage between DNA markers on the X chromosome and male sexual orientation', *Science*, 261 (1993), 321–7

70 Originally published in J. Brockman (ed.), *The Next Fifty Years* (New York, Vintage Books, 2002)

71 S. Brenner, 'Theoretical Biology in the Third Millennium', *Phil. Trans. Roy. Soc. B*, 354 (1999), 1963–5

72 *The Great Ape Project*, 25.

73 D. Dennett, *Consciousness Explained* (Boston, Little Brown, 1990). D. Dennett, *Darwin's Dangerous Idea* (New York, Simon & Schuster, 1995)

74 Foreword to S. Blackmore, *The Meme Machine* (Oxford, Oxford University Press, 1999)

75 J. D. Delius, 'The Nature of Culture' in M. S. Dawkins, T. R. Halliday and R. Dawkins (eds.), *The Tinbergen Legacy* (London, Chapman & Hall, 1991)

76 'Culturgen' was proposed by C. J. Lumsden and E. O. Wilson in *Genes, Mind and Culture* (Cambridge, Mass., Harvard University Press, 1981). Completely unknown to me when I coined 'meme' in 1976, the German biologist Richard Semon wrote a book called *Die Mneme* (English translation *The Mneme* (London, Allen & Unwin, 1921)) in which he adopted the 'mneme' coined in 1870 by the Austrian physiologist Ewald Hering. I first learned of this in a review of *The Selfish Gene* by Peter Medawar, who described the 'mneme' as 'a word of conscious etymological rectitude'

77 Originally published in B. Dahlbom (ed.), *Dennett and His Critics: Demystifying Mind* (Oxford, Blackwell, 1993)

78 D. Dennett, *Consciousness Explained* (Boston, Little Brown, 1990), p. 207

79 H. Thimbleby, 'Can viruses ever be useful?', *Computers and Security*, 10 (1991), 111–14

80 Sir Thomas Browne, *Religio Medici* (1635), I, 9

81 A. Zahavi, 'Mate selection – a selection for a handicap', *Journal of Theoretical Biology*, 53 (1975), 205–14

[82] A. Grafen, 'Sexual selection unhandicapped by the Fisher process', *Journal of Theoretical Biology*, 144 (1990), 473–516. A. Grafen, 'Biological signals as handicaps', *Journal of Theoretical Biology*, 144 (1990), 517–46

[83] M. Kilduff and R. Javers, *The Suicide Cult* (New York, Bantam, 1978)

[84] A. Kenny, *A Path from Rome* (Oxford, Oxford University Press, 1986)

[85] First published as 'Snake Oil and Holy Water' in *Forbes ASAP*, 4 October 1999

[86] U. Goodenough, *The Sacred Depths of Nature* (New York, Oxford University Press Inc., 1999)

[87] C. Sagan, *Pale Blue Dot: A Vision of the Human Future in Space* (New York, Ballantine, 1997)

[88] V. J. Stenger, *The Unconscious Quantum* (Buffalo, NY, Prometheus Books, 1996)

[89] The 'separate magisteria' thesis was promoted by S. J. Gould, an atheist bending over backwards far beyond the call of duty or sense, in *Rocks of Ages: science and religion in the fullness of life* (New York, Ballantine, 1999)

[90] First published in *The Independent*, 8 March 1997

[91] Originally published in *Freethought Today* (Madison, Wis.), 18: 8 (2001) (http://www.ffrf.org/). The text was revised for a special 'After Manhattan' edition of *The New Humanist* (Winter 2001)

[92] http://www.biota.org/people/douglasadams/index.html

[93] See also the splendid article by Polly Toynbee in *The Guardian* of 5 October 2001, http://guardian.co.uk/Columnists/Column/0,5673,563618,00.html

[94] http://www.guardian.co.uk/Archive/Article/0,4273,4257777,00.html

[95] W. D. Hamilton, *Narrow Roads of Gene Land*, vol. 2 (Oxford, Oxford University Press, 2001)

[96] John Diamond, *C: Because cowards get cancer too* (London, Vermilion, 1998)

[97] Published in *The Guardian*, 14 May 2001

[98] The full text of his speech may be seen at http://www.biota.org/people/douglasadams/index.html

[99] http://www.americanatheist.org/win98-99/T2/silverman.html

[100] *Break the Science Barrier with Richard Dawkins*, Channel 4, Equinox Series, 1996

[101] *Times Literary Supplement*, 11 September 1992. Originally published in Japanese as 'My Intended Burial and Why', *Insectarium*, 28 (1991), 238–47. Reprinted in English under the same title in *Ethology, Ecology & Evolution*, 12 (2000), 111–22

[102] W. D. Hamilton, 'Innate social aptitudes of man: an approach from evolutionary genetics' in R. Fox (ed.), *Biosocial Anthropology* (London, Malaby Press, 1975)

[103] W. D. Hamilton, *Narrow Roads of Gene Land*, vol. 1: Evolution of Social

Behaviour (Oxford, W. H. Freeman and Stockton Press, 1996). Volume 2 (Evolution of Sex) has now appeared (Oxford, Oxford University Press, 2001), with this eulogy as its Foreword

[104] John Diamond, *Snake Oil and Other Preoccupations* (London, Vintage, 2001)

[105] K. Sterelny, *Dawkins vs Gould: Survival of the Fittest* (Cambridge, Icon Books, 2001)

[106] A. Brown, *The Darwin Wars: How Stupid Genes Became Selfish Gods* (London, Pocket Books, 2000)

[107] *Lays of Ancient Rome*

[108] S. J. Gould, 'Self-help for a hedgehog stuck on a molehill' (review of R. Dawkins, *Climbing Mount Improbable*), *Evolution*, 51 (1997), 1020–3

[109] S. J. Gould, 'The Pattern of Life's History' in J. Brockman (ed.), *The Third Culture* (New York, Simon & Schuster, 1995), p. 64

[110] P. B. Medawar, *Art of the Soluble* (London, Penguin, 1969)

[111] Review of S. J. Gould, *Ever Since Darwin: Reflections in Natural History* (London, André Deutsch, 1978). First published in *Nature*, 276 (9 November 1978), 121–3

[112] Reprinted as 'Caring Groups and Selfish Genes' in S. J. Gould, *The Panda's Thumb* (New York, W. W. Norton, 1980)

[113] G. C. Williams, *Adaptation and Natural Selection* (Princeton, Princeton University Press, 1966), pp. 22–5 and 56–7

[114] P. B. Medawar, *Pluto's Republic* (New York, Oxford University Press Inc., 1982)

[115] S. J. Gould, *Hen's Teeth and Horse's Toes* (New York, W. W. Norton, 1983)

[116] P. B. Medawar, *The Hope of Progress* (London, Methuen, 1972)

[117] R. Dawkins, *The Selfish Gene*, 2nd edn (Oxford, Oxford University Press, 1989), pp. 271–2. See also R. Dawkins, *The Extended Phenotype* (Oxford University Press, 1999), pp. 116–17, 239–47

[118] Review of S. J. Gould, *Wonderful Life* (London, Hutchinson Radius, 1989). Published in the *Sunday Telegraph*, 25 February 1990

[119] *Daily Telegraph*, 22 January 1990

[120] Review of S. J. Gould, *Full House* (New York, Harmony Books, 1996); published in the UK as *Life's Grandeur* (London, Jonathan Cape, 1996). In *Evolution*, 51:3 June 1997), 1015–20

[121] I have devoted a whole article to attacking the idea of progress in this sense: R. Dawkins, 'Progress' in E. Fox Keller and E. Lloyd (eds.), *Keywords in evolutionary biology* (Cambridge, Mass., Harvard University Press, 1992), pp. 263–72

[122] J. Maynard Smith, 'Time in the Evolutionary Process', *Studium Generale*, 23 (1970), 266–72

[123] D. W. McShea, 'Metazoan complexity and evolution: is there a trend?', *Evolution*, 50 (1996), 477–92

[124] J. W. S. Pringle, 'On the parallel between learning and evolution', *Behaviour*, 3 (1951), 90–110

[125] J. Huxley, *The Individual in the Animal Kingdom* (Cambridge, Cambridge University Press, 1912)

[126] J. Huxley, *Essays of a Biologist* (London, Chatto & Windus, 1926)

[127] S. Pinker, *The Language Instinct* (London, Viking, 1994)

[128] M. Ridley, 'Coadaptation and the inadequacy of natural selection', *Brit. J. Hist. Sci.*, 15 (1982), 45–68

[129] R. Dawkins and J. R. Krebs, 'Arms races between and within species', *Proc. Roy. Soc. Lond.* B, 205 (1979), 489–511

[130] H. Jerison, *Evolution of the brain and intelligence* (New York, Academic Press, 1973)

[131] J. Maynard Smith, 'Genes, Memes and Minds', *New York Review of Books*, 30 (30 November 1995). Review of D. Dennett, *Darwin's Dangerous Idea*

[132] R. Leakey and R. Lewin, *The Sixth Extinction* (London, Weidenfeld & Nicolson, 1996)

[133] G. A. Wray, J. S. Levinton and L. H. Shapiro, 'Molecular Evidence for Deep Precambrian Divergences Among Metazoan Phyla', *Science* 274 (1996), 568

[134] http://www.arn.org/docs/pjweekly/pj_weekly_011202.htm

[135] S. J. Gould, *The Structure of Evolutionary Theory* (Cambridge, Mass., Harvard University Press, 2002)

[136] D. Barash, 'Grappling with the Ghost of Gould', *Human Nature Review*, 2 (9 July 2002), 283–92

[137] Foreword to H. Croze and J. Reader, *Pyramids of Life* (London, Harvill Press, 2000)

[138] Originally published as an article on E. Huxley, *Red Strangers* (London, Chatto, 1964) in the *Financial Times*, 9 May 1998; later as Foreword to the book, republished by Penguin Books (1999)

[139] Angus, Maisie and Travers McNeice, *The Lion Children* (London, Orion Books, 2001)

[140] First published as 'All Our Yesterdays' in the *Sunday Times*, 31 December 1995

[141] R. Leakey, *The Origin of Humankind* (London, Weidenfeld & Nicolson, 1994)

[142] I. Douglas-Hamilton and O. Douglas-Hamilton, *Among the Elephants* (London, Viking, 1975). I. Douglas-Hamilton and O. Douglas-Hamilton, *Battle for the Elephants* (London, Doubleday, 1992)

[143] P. Matthiessen, *The Tree where Man was Born* (London, Harvill Press, 1998)

[144] Published in J. Brockman and K. Matson (eds.), *How Things Are* (New York, Morrow, 1995)

TEXT ACKNOWLEDGEMENTS

The publishers wish to thank the individuals and organizations who have given their permission to reproduce the essays listed below. Every effort has been made to contact and receive permission from the copyright owners to reproduce the essays in this volume. Any errors, oversights or omissions will be corrected in subsequent editions.

'What is True?', originally published as 'Hall of Mirrors' in *Forbes ASAP*, reprinted by permission of *Forbes ASAP* Magazine © 2002 Forbes Inc.; 'Postmodernism Disrobed', reprinted by permission from *Nature*, 394 (1998), 141–3, © 1998 Macmillan Publishers Ltd.; 'Light Will Be Thrown', originally published as the Foreword to *The Descent of Man: A New Edition* (Gibson Square Books, 2003), reprinted by permission of Gibson Square Books Ltd.; 'Darwin Triumphant', from M. H. Robinson and L. Tiger (eds), *Man and Beast Revisited* (Smithsonian Institution Press, 1991), reprinted by permission of the Smithsonian Institution Press; 'The "Information Challenge"', reprinted by permission from *the Skeptic*, 18:4 (December 1998); 'Son of Moore's Law', from J. Brockman (ed.), *The Next Fifty Years* (Weidenfeld & Nicolson, 2002), reprinted by permission of The Orion Publishing Group and John Brockman; 'Chinese Junk and Chinese Whispers', originally published as the Foreword to S. Blackmore, *The Meme Machine* (Oxford University Press, 1999), reprinted by permission of Oxford University Press; 'Viruses of the Mind', originally published in B. Dahlbom (ed.), *Dennett and His Critics: Demystifying Mind* (Blackwell, 1993), reprinted by permission of Blackwell Publishers; 'The Great Convergence', originally published as 'Snake Oil and Holy Water' in *Forbes ASAP*, reprinted by permission of *Forbes ASAP* Magazine © 2002 Forbes Inc.; 'Rejoicing in Multifarious Nature', reprinted by permission from *Nature*, 276 (1978), 121–3, © 1978 Macmillan Publishers Ltd.; 'Human Chauvinism', reprinted by permission from *Evolution*, 51:3 (June 1997), 1015–20; 'I Speak of Africa and Golden Joys', originally published as the Foreword to A. McNeice, M. McNeice and T. McNeice, *The Lion Children* (Orion Books, 2001), reprinted by permission of The Orion Publishing Group; 'Good and Bad Reasons for Believing', originally published in J. Brockman and K. Matson (eds), *How Things Are: A Science Tool-kit for the Mind* (Morrow, 1995), reprinted by permission of John Brockman.

INDEX

Abortion, *see* Ethics
Acquired characteristics, Inheritance
    of, *see* Lamarckian theory
Adams, Douglas, 164, 184–185, 186,
    191–192, 193–199
Adaptations
    As state of order, 99
    Ecosystems are not, 266
    Lamarckian cannot explain, 105
    Only cumulative evolution builds
      complex, 102, 250
    Result of nonrandom guiding
      mechanism, 103
Adaptive radiation, 252
Africa, 263–264, 265, 268–269,
    271–282
Agassiz, Louis, 232
Ageing
    Medawar/Williams theory of,
      155
    W. D. Hamilton on, 202, 207
Agnostic conciliation, 176–177
Alpha globin, *see* Globin
'Alternative' medicine, 41, 42, 192,
    209–215
Amino acid, 114, 133
Ammophila, *see* Digger wasp
Ancestor, Common, 119
    Of all living species, 77, 113
    Of apes, 26–27
    Of Cambrian phyla, 252, 253, 254
    Of humans and chimpanzees,
      26–29, 86, 135
    Of vertebrates, 115–116
Angier, Natalie, 200
Archimedes, 94f
Ardipithecus, 86

Assumption, Doctrine of, 179, 286,
    287–288
Astrology, 8, 50
Aunger, Robert, 150f
Australopithecus, 28, 86, 87, 135, 281
Autumn leaves, Hamilton on, 202
Axelrod, Robert, 202
Axolotl, 87

Bacteria
    'Age of', 246
    Animal as community of, 267
    Antibiotic resistance among, 34
Balkin, J. M., 150f
Barash, David, 261
Barlow, Horace, 111f
Bartz, Stephen, 201–202
Bateson, William, 93
Baudrillard, Jean, 59
Bauplan, 254
Benedict XIV, 179
Bentham, Jeremy, 29
Beta globin, *see* Globin
Bipedality, 87
Birch, Martin, 206
Bishop of Oxford, 94f
Bit, 109, 110
Blackmore, Susan, 137, 144, 150f
Blair, Tony, 8, 31, 184
Blind Watchmaker, The, 13, 97f
Bloom, H., 150f
Blueprint vs. Recipe, 104, 124
Boswell, James, 209
Boyer, Pascal, 138f
Bozzi, Luisa, 203
Bradman, Sir Donald, 31
Brain, Human